Homework Book, Hig

Delivering the Edexcel Specification

NEW GCSE MATHS
Edexcel Modular

Fully supports the 2010 GCSE Specification

**Brian Speed • Keith Gordon • Kevin Evans
Trevor Senior • Chris Pearce**

CONTENTS

CORE

UNIT 1: Statistics and Number

UNIT 2: Number, Algebra and Geometry

INTRODUCTION

Welcome to Collins New GCSE Maths for Edexcel Modular Higher Homework Book 1. This book follows the structure of the Edexcel Modular Higher Student Book and provides homework questions to cover topics in Units 1 and 2.

Colour-coded grades

Know what target grade you are working at and track your progress with the colour-coded grade panels at the side of the page.

Use of calculators

Questions for which you could use a calculator are marked with a  icon. Remember in your Unit 2 exam you will not be allowed to use a calculator.

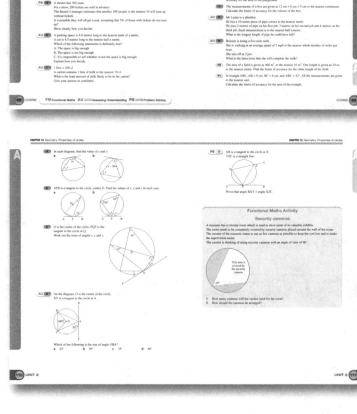

Examples

Recap on methods you need by reading through the examples before starting the homework exercises.

Functional maths

Practise functional maths skills to see how people use maths in everyday life. Look out for practice questions marked **FM**.

There are also extra functional maths and problem-solving activities at the end of every chapter to build and apply your skills.

New Assessment Objectives

Practise new parts of the curriculum (Assessment Objectives AO2 and AO3) with questions that assess your understanding marked **AU** and questions that test if you can solve problems marked **PS**. You will also practise some questions that involve several steps and where you have to choose which method to use; these also test AO2. There are also plenty of straightforward questions (AO1) that test if you can do the maths.

Student Book CD-ROM

Remind yourself of the work covered in class with the Student Book in electronic form on the CD-ROM. Insert the CD into your machine and choose the chapter you need.

1 Number: Number skills and properties

1.1 Solving real-life problems

HOMEWORK 1A

FM 1 Andy needs enough tiles to cover 12 m^2 in his bathroom. It takes 25 tiles to cover 1 m^2. It is recommended to buy 20 percent more tiles to allow for cutting. Tiles are sold in boxes of 16. Andy buys 24 boxes. Does he have enough tiles?

FM 2 The organiser of a church fête needs 1000 balloons. She has a budget of £30. Each packet contains 25 balloons and costs 85p. Does she have enough?

FM 3 A TV rental shop buys TVs for £110 each. The shop needs to make a least 10 percent profit on each TV to cover its costs. On average, each TV is rented for 40 weeks at £3.50 per week. Does the shop cover its costs?

4 The annual subscription fee to join a fishing club is £42. The treasurer of the club has collected £1134 in fees. How many people have paid their subscription fee?

5 Mrs Woodhead saves £14 per week towards her bills. How much does she save in a year?

FM 6 Mark saves £15 each week. He wants to buy a dining set costing £860. Will he have saved enough money after one year? Show how you worked out your answer.

7 Sylvia has a part-time job and is paid £18 for every day she works. Last year she worked for 148 days. How much was she paid for the year?

PS 8 Mutya has a part-time job, working three days each week. She is paid £5 per hour. She works 4 hours each day. Neil has a full-time job, working five days each week. He is paid £6 per hour. He works 7 hours each day. How many weeks does Mutya have to work to earn at least as much as Neil earns in one week? Show how you worked out your answer.

FM 9 A coach firm charges £504 for 36 people to go Christmas shopping on a day trip to Calais. The cost of the coach is shared equally between the passengers. Mary takes £150. When she gets to Calais, Mary wants to buy games costing €50 each for each of her four grandchildren. The exchange rate is £1 = €1.25. Does she have enough money to pay for the presents?

10 A concert hall has 48 rows of seats with 32 seats in a row. What is the maximum capacity of the hall?

AU 11 Allan is a market gardener and has 420 bulbs to plant. He plants them in rows with 18 bulbs to a row. How many complete rows will there be?

FM 12 A room measuring 6 m by 8 m is to be carpeted. The carpet costs £19 per m².
- **a** Estimate the cost of the carpet.
- **b** Calculate the exact cost of the carpet.

AU 13 Paul's room measures 7 m by 3 m.
The carpet is sold from rolls which are 4 m wide.
This means that a person buying carpet has to buy pieces from the full width of the roll.
- **a** What is the smallest area of carpet that Paul has to buy?
- **b** He has £300 to spend on the carpet.
  What is the most he can spend per m²?

FM 14 There are 240 students and teachers on a school visit.
Four coaches are booked.
Each coach can take 53 passengers.
The school decides to book another coach for the remaining passengers.
What is the smallest number of seats needed on this coach?

PS 15 On average, a toy shop sells 38 computer games each week.
The manager has a delivery of 150 games each month.
Will his stock of games increase or decrease?
Show clearly how you decide.

1.2 Approximation of calculations

HOMEWORK 1B

1 Round each of the following to one significant figure.
a	51 203	**b**	56 189	**c**	33 261	**d**	89 998	**e**	94 999
f	53.71	**g**	87.24	**h**	31.06	**i**	97.835	**j**	184.23
k	0.5124	**l**	0.2765	**m**	0.006 12	**n**	0.049 21	**o**	0.000 888
p	9.7	**q**	85.1	**r**	91.86	**s**	196	**t**	987.65

AU 2 What is the least and the greatest number of people that can be found in these towns?
Hellaby	population 900 (to 1 significant figure)
Hook	population 650 (to 2 significant figures)
Hundleton	population 1050 (to 3 significant figures)

3 Round off each of the following numbers to 2 significant figures.
a	6725	**b**	35 724	**c**	68 522	**d**	41 689	**e**	27 308
f	6973	**g**	2174	**h**	958	**i**	439	**j**	327.6

4 Round off each of the following to the number of significant figures (sf) indicated.
a	46 302 (1 sf)	**b**	6177 (2 sf)	**c**	89.67 (3 sf)	**d**	216.7 (2 sf)
e	7.78 (1 sf)	**f**	1.087 (2 sf)	**g**	729.9 (3 sf)	**h**	5821 (1 sf)
i	66.51 (2 sf)	**j**	5.986 (1 sf)	**k**	7.552 (1 sf)	**l**	9.7454 (3 sf)
m	25.76 (2 sf)	**n**	28.53 (1 sf)	**o**	869.89 (3 sf)	**p**	35.88 (1 sf)
q	0.084 71 (2 sf)	**r**	0.0099 (2 sf)	**s**	0.0809 (1 sf)	**t**	0.061 97 (3 sf)

PS 5 A baker estimates that she has baked 100 loaves. She is correct to one significant figure.
She sells two loaves and now has 90 loaves to one significant figure.
How many could she have had to start with?
Work out all possible answers.

AU 6 There are 500 cars in a car park to one significant figure.
What is the least possible number of cars that could enter the car park so that there are 700 cars in the car park to one significant figure?

FM 7 A supermarket manager estimates that five minutes before closing time there are 200 people still shopping.
This number is to one significant figure.
On average a checkout serves four customers in five minutes.
No more shoppers enter the supermarket.
How many checkouts should he open to be sure to serve all the shoppers by closing time?

HOMEWORK 1C

1 Without using a calculator, write down the answer to these.
    a 50×600 **b** 0.6×40 **c** 0.02×400 **d** $(30)^2$
    e 0.5×250 **f** 0.6×0.7 **g** $30 \times 40 \times 50$ **h** $200 \times 0.7 \times 40$

2 Without using a calculator, write down the answer to these.
    a $4000 \div 20$ **b** $8000 \div 200$ **c** $400 \div 0.5$ **d** $2000 \div 0.05$
    e $1800 \div 0.12$ **f** $600 \div 0.3$ **g** $200 \times 30 \div 40$ **h** $300 \times 70 \div 0.4$

AU 3 You are given that $18 \times 21 = 378$
Write down the value of:
    a 180×210
    b $3780 \div 21$

PS 4 Match each calculation to its answer and then write out the calculations in order, starting with the smallest answer.
    6000×300 500×7000 $10\,000 \times 900$ $20 \times 80\,000$
    $3\,500\,000$ $1\,800\,000$ $1\,600\,000$ $9\,000\,000$

5 The Moon is approximately 400 000 km from Earth.
If a spaceship takes 8 days to reach the Moon and return, how far does it travel each day?

HOMEWORK 1D

1 Find approximate answers to the following sums.
    a 4324×6.71 **b** 6170×7.311 **c** 72.35×3.142
    d 4709×3.81 **e** $63.1 \times 4.18 \times 8.32$ **f** $320 \times 6.95 \times 0.98$
    g $454 \div 89.3$ **h** $26.8 \div 2.97$ **i** $4964 \div 7.23$
    j $316 \div 3.87$ **k** $2489 \div 48.58$ **l** $63.94 \div 8.302$

2 Find the approximate monthly pay of the following people whose annual salary is:
    a Joy £47 200   **b** Amy £24 200   **c** Tom £19 135

3 Find the approximate annual pay of these brothers who earn:
    a Trevor £570 a week   **b** Brian £2728 a month

AU 4 A groundsman bought 350 kg of seed at a cost of £3.84 per kg. Find the approximate total cost of this seed.

FM 5 A greengrocer sells a box of 250 apples for £47. He knows that if he has sold them for 20p each or more he will make a profit.
Did he make a profit? Explain why using approximations.

6 Keith runs about 15 km every day. Approximately how far does he run in:
    a a week   **b** a month   **c** a year?

7 A litre of creosote will cover an area of about 6.8 m². Approximately how many litre cans will I need to buy to creosote a fence with a total surface area of 43 m²?

PS 8 A tour of London sets off at 10.13 am and costs £21. It returns at 12.08 pm. Approximately how much is the tour per hour?

HOMEWORK 1E

FM 1 Round each of the following to give sensible answers.
   a Kris is 1.6248 m tall.
   b It took me 17 minutes 48.78 seconds to cook the dinner.
   c My rabbit weighs 2.867 kg.
   d The temperature at the bottom of the ocean is 1.239 °C.
   e There were 23 736 people at the game yesterday.

2 How many jars each holding 119 cm³ of water can be filled from a 3-litre flask?

AU 3 If I walk at an average speed of 62 m per minute, how long will it take me to walk a distance of 4 km?

FM 4 Helen earns £31 500 a year. She works 5 days a week for 45 weeks a year. How much does she earn a day?

5 10 g of gold costs £2.17. How much will 1 kg of gold cost?

FM 6 Rewrite the following article using sensible numbers.
I left home at eleven and a half minutes past two, and walked for 49 minutes. The temperature was 12.7623 °C. I could see an aeroplane overhead at 2937.1 feet. Altogether I had walked 3.126 miles.

PS 7 David travelled 350 miles in 5 hours 10 minutes.
Trevor travelled half the distance in half the time.
Approximately how fast was Trevor travelling?

1.3 Negative numbers

HOMEWORK 1F

FM AU 1   **a** Work out $17 \times (-4)$
   b The average temperature drops by 4°C every day for 17 days. How much has the temperature dropped altogether?
   c The temperature drops by 6°C for the next four days. Write down the calculation to work out the total drop in temperature over these three days.

2 Write down the answers to the following.
a -2×4	**b** -3×6	**c** -5×7	**d** $-3 \times (-4)$	**e** $-8 \times (-2)$
f $-14 \div (-2)$	**g** $-16 \div (-4)$	**h** $25 \div (-5)$	**i** $-16 \div (-8)$	**j** $-8 \div (-4)$
k $3 \times (-7)$	**l** $6 \times (-3)$	**m** $7 \times (-4)$	**n** $-3 \times (-9)$	**o** $-7 \times (-2)$
p $28 \div (-4)$	**q** $12 \div (-3)$	**r** $-40 \div 8$	**s** $-15 \div (-3)$	**t** $50 \div (-2)$
u $-3 \times (-8)$	**v** $42 \div (-6)$	**w** $7 \times (-9)$	**x** $-24 \div (-4)$	**y** -7×8

3 Write down the answers to the following.

a $-2+4$	**b** $-3+6$	**c** $-5+7$	**d** $-3+(-4)$	**e** $-8+(-2)$
f $-14-(-2)$	**g** $-16-(-4)$	**h** $25-(-5)$	**i** $-16-(-8)$	**j** $-8-(-4)$
k $3+(-7)$	**l** $6+(-3)$	**m** $7+(-4)$	**n** $-3+(-9)$	**o** $-7+(-2)$
p $28-(-4)$	**q** $12-(-3)$	**r** $-40-8$	**s** $-15-(-3)$	**t** $50-(-2)$
u $-3+(-8)$	**v** $42-(-6)$	**w** $7+(-8)$	**x** $-24-(-4)$	**y** $-7+8$

4 What number do you multiply -5 by to get the following?

a 25 **b** -30 **c** 50 **d** -100 **e** 75

PS 5 Put these calculations in order from lowest to highest.

$-18 \div 12$ $-0.5 \times (-4)$ $-21 \div (-14)$ $0.3 \times (-2)$

HOMEWORK 1G

1 Work out each of these. Remember: first work out the bracket.

a $-3 \times (-2+6)$ **b** $8 \div (-3+2)$ **c** $(6-8) \times (-3)$ **d** $-4 \times (-6-3)$
e $-5 \times (-6 \div 2)$ **f** $(-5+3) \times (-3)$ **g** $(6-9) \times (-4)$ **h** $(2-5) \times (5-2)$

2 Work out each of these.

a $-5 \times (-4)+3$ **b** $-8 \div 8-3$ **c** $16 \div (-4)+3$ **d** $2 \times (-5)+6$
e $-3 \times 4-5$ **f** $-1+4^2-5$ **g** $5-3^2+2$ **h** $-1+2 \times (-3)$

PS 3 Copy each of these and then put in a bracket to make each one true.

a $4 \times (-3)+2=-4$ **b** $-6 \div (-3)+2=4$ **c** $-6 \div (-3)+2=6$

4 $a=-3, b=5, c=-4$

Work out the values of the following.

a $(a+b)^2$ **b** $-(a+c)^2$ **c** $(a+b)c$ **d** $a^2+b^2+c^2$

5 Work out each of the following.

a $(7^2+2^2) \times 3$
b $18 \div (2-5)^2$
c $3 \times (6^2-(1-8)^2)$
d $((7+1)^2-(2-3)^2) \div 7$

AU 6 Use each of the numbers 4, 6 and 8 and each of the symbols $-$, $\times$ and $\div$ to make a calculation with an answer -3.

AU 7 Use any four different numbers to make a calculation with an answer of -9.

PS 8 Use the numbers 1, 2, 3, 4 and 5, in order from smallest to largest, together with one of each of the symbols $+$, $-$, $\times$, $\div$ and two pairs of brackets to make a calculation with an answer of -2.75.

For example: Making a calculation with an answer of -4:

$(1+2)-(3 \times 4)+5=-4$

Functional Maths Activity

Carpet bargains

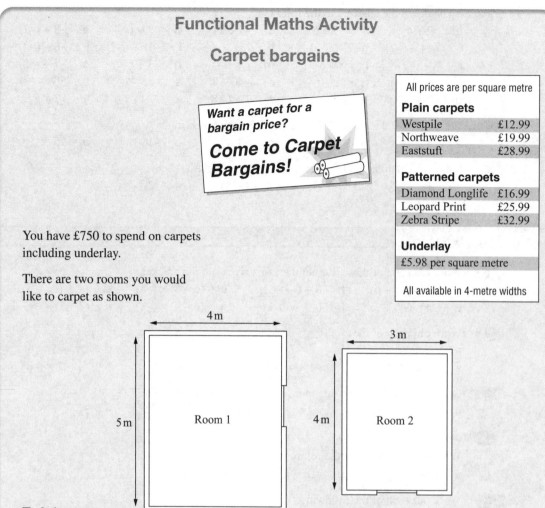

Want a carpet for a bargain price?

Come to Carpet Bargains!

All prices are per square metre	
Plain carpets	
Westpile	£12.99
Northweave	£19.99
Eaststuft	£28.99
Patterned carpets	
Diamond Longlife	£16.99
Leopard Print	£25.99
Zebra Stripe	£32.99
Underlay	
£5.98 per square metre	

All available in 4-metre widths

You have £750 to spend on carpets including underlay.

There are two rooms you would like to carpet as shown.

Room 1 — 4 m × 5 m

Room 2 — 3 m × 4 m

Task 1

Work out the cost of buying carpet and underlay for Room 1 using Diamond Longlife.

Task 2

You have £750 to spend on carpets for the two rooms shown here including underlay.
You want to buy the best carpets that you can afford.
You may choose to carpet one room or two rooms.
Choose which carpets to fit in the rooms and work out the costs. Make sure you stay within your budget of £750.

Task 3

There is a special offer.
Repeat task 2 using the offer.

Special offers throughout the store

Half marked price
plus all carpets fitted free*

*Free fitting when you buy the equivalent size of underlay

Number: Fractions, percentages and ratios

2.1 One quantity as a fraction of another

HOMEWORK 2A

1 Write the first quantity as a fraction of the second.
  a 5p, 20p   **b** 4 kg, 12 kg   **c** 6 hours, 12 hours
  d 14 days, 30 days   **e** 6 days, 2 weeks   **f** 20 minutes, 2 hours

2 David scored 18 goals out of his team's season total of 48 goals. What fraction of all the goals did David score?

3 James works for 32 weeks a year. What fraction of the year does he work for?

FM 4 Mark earns £120 and saves £40 of it.
Bev earns £150 and saves £60 of it.
Who is saving the greater proportion of their earnings?

FM 5 In a test, Kevin scores 7 out of 10 and Sally scores 15 out of 20. Which is the better mark?
Explain your answer.

AU 6 In a dancing group, there are 20 dancers, of whom 12 are women. Half of the men are single.
What fraction of the dancers are single men?
Give your answer in its simplest form.

PS 7 A bus driver says that at least two out of every three passengers on her bus have a travel pass.
A different driver says it is less than three out of four.
If both statements are true and the bus is carrying 50 passengers, how many have a travel pass? Write down all possible answers.

2.2 Increasing and decreasing quantities by a percentage

HOMEWORK 2B

1 Increase each of the following by the given amount.
  a £80 by 5%   **b** 14 kg by 6%   **c** £42 by 3%

2 Increase each of the following by the given amount.
  a 340 g by 10%   **b** 64 m by 5%   **c** £41 by 20%

3 Keith, who was on a salary of £34 200, was given a pay rise of 4%. What was his new salary?

4 In 1991 the population of Dripfield was 14 200. By 2001 that had increased by 8%. What was the population of Dripfield in 2001?

5 In 1993 the number of bikes on the roads of Doncaster was about 840. Since then it has increased by 8%. Approximately how many bikes are on the roads of Doncaster now?

PS **6** A dining table costs £300 before the VAT is added.
If the rate of VAT goes up from 15% to 20%, how much will the cost of the dining table increase?

FM **7** A restaurant meal is advertised at £20 (a service charge will be added to all bills).

Romano's Bistro

2 × Prawn Cocktail	4.60
1 × Chicken Risotto	6.60
1 × Sea Bass	8.50
1 × Fries	4.50
1 × Tiramisu	6.60
1 × Choc fudge cake	5.20
2 × Coffee	5.00
Service charge	5.00
Total	46.00

The bill for two people is shown.
Show that the service charge is 15%.

PS **8** A shopkeeper decides to increase prices by 5% or £1, whichever is greater.
What is the price range of the items that will rise by £1?

HOMEWORK 2C

 1 Decrease each of the following by the given percentage. (Use any method you like.)
 a £20 by 10%   **b** £150 by 20%   **c** 90 kg by 30%   **d** 500 m by 12%
 e £260 by 5%   **f** 80 cm by 25%   **g** 400 g by 42%   **h** £425 by 23%
 i 48 kg by 75%   **j** £63 by 37%

 FM **2** Mrs Denghali buys a new car from a garage for £8400. The garage owner tells her that the value of the car will lose 24% after one year. What will be the value of the car after one year?

 3 The population of a village in 2006 was 2400. In 2010 the population had decreased by 12%. What was the population of the village in 2010?

 FM **4** A Travel Agent is offering a 15% discount on holidays. How much will the advertised holiday now cost?

NEW YORK FOR A WEEK
£540

 FM **5**

New Year's Sale:
All prices reduced by 20%

Matt was given £160 at Christmas. Can he afford to buy a shirt that normally costs £30, a suit that normally costs £130 and a pair of shoes that normally cost £42?

6 On the first day of a new term, a school expects to have an attendance rate of 99%. If the school population is 700 pupils, how many pupils will the school expect to be absent on the first day of the new term?

 7 By putting cavity wall insulation into your home, you could use 20% less fuel. A family using an average of 850 units of electricity a year put cavity wall insulation into their home. How much electricity would they expect to use now?

AU PS 8 A shop increases all its prices by 10%.

 One month later it advertises 10% off all marked prices.
Are the goods cheaper, the same or more expensive than before the price increase?
Show how you work out your answer.

PS 9 Prove that a 10% increase followed by a 10% decrease is equivalent to a 1% increase overall.

2.3 Expressing one quantity as a percentage of another

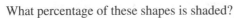 **HOMEWORK 2D**

1 Express each of the following as a percentage. Give your answers to one decimal place where necessary.
  a £8 of £40   **b** 20 kg of 80 kg   **c** 5 m of 50 m
  d £15 of £20   **e** 400 g of 500 g   **f** 23 cm of 50 cm
  g £12 of £36   **h** 18 minutes of 1 hour   **i** £27 of £40
  j 5 days of 3 weeks

PS 2 What percentage of these shapes is shaded?
  a   **b**

3 In a class of 30 pupils, 18 are girls.
  a What percentage of the class are girls?
  b What percentage of the class are boys?

4 The area of a farm is 820 hectares. The farmer uses 240 hectares for pasture.
  What percentage of the farm land is used for pasture? Give your answer to one decimal place.

5 Find the percentage profit on the following items. Give your answers to one decimal place.

Item (Selling price)	Retail price (Price the shop paid)	Wholesale price
Micro hi-fi system	£250	£150
CD radio cassette	£90	£60
MiniDisc player	£44.99	£30
Cordless headphones	£29.99	£18

AU 6 Paul and Val take the same tests. Both tests are worth the same number of marks.
Here are their results.

	Test A	Test B
Paul	30	40
Val	28	39

Whose result has the greater percentage increase from test A to test B?
Show your working.

PS 7 A small train is carrying 48 passengers.
At a station, more passengers get on so that all seats are filled and no one is standing.
At the next station, 70% of the passengers leave the train and 30 new passengers get on.
There are now 48 passengers on the train again.
How many seats are on the train?

PS 8 In a secondary school, 30% of students have a younger brother or sister at primary
school.
20% of students have two younger brothers or sisters at primary school
Altogether there are 700 brothers and sisters at the primary school.
How many students are at the secondary school?

FM 9 James came home from school one day with his end-of-year test results. Change each of
James' results to a percentage.

| Maths | 63 out of 75 | English | 56 out of 80 |
| Science | 75 out of 120 | French | 27 out of 60 |

10 During the wet year of 1993, it rained in London on 80 days of the year. What percentage
of the days of the year were wet? Give your answer to two significant figures.

2.4 Compound interest and repeated percentage change

HOMEWORK 2E

1 A small plant increases its height by 10% each day in the second week of its growth.
At the end of the first week, the plant was 5 cm high.
What is its height after a further:
  a 1 day **b** 2 days **c** 4 days **d** 1 week?

FM 2 The headmaster of a new school offered his staff an annual pay increase of 5% for every
year they stayed with the school.
  a Mr Speed started teaching at the school on a salary of £28 000. What salary will he
  be on after 3 years if he stays at the school?
  b Miss Tuck started teaching at the school on a salary of £14 500. How many years
  will it be until she is earning a salary of over £20 000?

3 Billy put a gift of £250 into a special savings account that offered him 8% compound
interest if he promised to keep the money in for at least 2 years. How much was in this
account after:
  a 2 years **b** 3 years **c** 5 years?

4 The penguin population of a small island was only 1500 in 1998, but it steadily increased
by about 15% each year. Calculate the population in:
  a 1999 **b** 2000 **c** 2002

PS 5 A sycamore tree is 40 cm tall; it grows at a rate of 8% per year. A conifer is 20 cm tall. It grows at a rate of 15% per year. How many years does it take before the conifer is taller than the sycamore?

AU 6 The population of a small town is 2000.
The population is falling by 10% each year.
The population of a nearby village is 1500.
This population is rising by 10% each year.
After how many years will the population of the town be less than the population of the village?

PS 7 Each week, a boy takes out 20% of the amount in his bank account to spend.
After how many weeks will the amount in his bank account have halved from the original amount?

2.5 Reverse percentage (working out the original quantity)

 HOMEWORK 2F

1 Find what 100% represents when:
  a 20% represents 160 g   **b** 25% represents 24 m   **c** 5% represents 42 cm.

2 Find what 100% represents when:
  a 40% represents 28 kg   **b** 30% represents £54   **c** 15% represents 6 hours.

3 VAT is a government tax added to goods and services. With VAT at 17.5%, what is the pre-VAT price of the following priced goods?
  Jumper   £14.10   Socks   £1.88   Trousers   £23.50

4 Paula spends £9 each week on CDs. This is 60% of her weekly income. How much is Paula's weekly income?

5 Alan's weekly pay is increased by 4% to £187.20. What was Alan's pay before the increase?

FM 6 Jon's salary is now £23 980. This is 10% more than he earned two years ago.
Last year his salary was 3% more than it was two years ago.
  a How much was his salary last year?
  b By what percentage was his salary increased last year?

PS AU 7 Twice as many people visit a shopping centre on Saturdays than on Fridays.

The number visiting on both days increases by 50% in the week before Christmas. How many more visit on this Saturday than on this Friday? Give your answer as a percentage.

PS 8 A man's savings decreased by 10% in one year and then increased in the following year by 10%.
He now has £1782.
How much did he have two years ago?

2.6 Ratio

HOMEWORK 2G

1 Express each of the following ratios in their simplest form.
   a 3 : 9   **b** 5 : 25   **c** 4 : 24   **d** 10 : 30   **e** 6 : 9
   f 12 : 20   **g** 25 : 40   **h** 30 : 4   **i** 14 : 35   **j** 125 : 50

2 Express each of the following ratios of quantities in their simplest form. (Remember to change to a common unit where necessary.)
   a £2 to £8   **b** £12 to £16   **c** 25 g to 200 g
   d 6 miles : 15 miles   **e** 20 cm : 50 cm   **f** 80p : £1.50
   g 1 kg : 300g   **h** 40 seconds : 2 minutes   **i** 9 hours : 1 day
   j 4 mm : 2 cm

3 £20 is shared out between Bob and Kathryn in the ratio 1 : 3.
   a What fraction of the £20 does Bob receive?
   b What fraction of the £20 does Kathryn receive?

4 In a class of students, the ratio of boys to girls is 2 : 3.
   a What fraction of the class is boys?
   b What fraction of the class is girls?

FM 5 Pewter is an alloy containing lead and tin in the ratio 1 : 9.
   a What fraction of pewter is lead?
   b What fraction of pewter is tin?

AU 6 Roy wins two-thirds of his snooker matches. He loses the rest.
   What is his ratio of wins to losses?

PS 7 In the 2009 Ashes cricket series, the numbers of wickets taken by Steve Harmison and Monty Panesar were in the ratio 5 : 1.
   The ratio of the number of wickets taken by Graham Onions to those taken by Steve Harmison was 2 : 1.
   What fraction of the wickets taken by these three bowlers was by Monty Panesar?

FM 8 In a motor sales business, the ratio of the area allocated to selling new cars to the area for selling used cars is 3 : 2.
   Half the area for selling new cars is changed to used cars.
   What is the ratio of the areas now?
   Give your answer in its simplest form.

HOMEWORK 2H

1 Divide each of the following amounts in the given ratios.
   a £10 in the ratio 1 : 4   **b** £12 in the ratio 1 : 2
   c £40 in the ratio 1 : 3   **d** 60 g in the ratio 1 : 5
   e 10 hours in the ratio 1 : 9

2 The ratio of female to male members of a sports centre is 3 : 1. The total number of members of the centre is 400.
   a How many members are female?   **b** How many members are male?

3 A 20 m length of cloth is cut into two pieces in the ratio 1 : 9. How long is each piece?

4 Divide each of the following amounts in the given ratios.
   a   25 kg in the ratio 2 : 3   **b**   30 days in the ratio 3 : 2
   c   70 m in the ratio 3 : 4   **d**   £5 in the ratio 3 : 7
   e   1 day in the ratio 5 : 3

5 James collects beer mats and the ratio of British mats to foreign mats in his collection is 5 : 2. He has 1400 beer mats. How many foreign beer mats does he have?

6 Patrick and Jane share out a box of sweets in the ratio of their ages. Patrick is 9 years old and Jane is 11 years old. If there are 100 sweets in the box, how many does Patrick get?

AU 7 For her birthday Reena is given £30. She decides to spend four times as much as she saves. How much does she save?

8 Mrs Megson calculates that her quarterly electric and gas bills are in the ratio 5 : 6. The total she pays for both bills is £66. How much is each bill?

9 You can simplify a ratio by changing it into the form 1 : n. For example, 5 : 7 can be rewritten as 5 : 7 = 1 : 1.4 by dividing each side of the ratio by 5. Rewrite each of the following ratios in the form 1 : n.
   a   2 : 3   **b**   2 : 5   **c**   4 : 5   **d**   5 : 8   **e**   10 : 21

PS 10 The amount of petrol and diesel sold at a garage is in the ratio 2 : 1. One-tenth of the diesel sold is bio-diesel.
What fraction of all the fuel sold is bio-diesel?

PS FM 11 At a buffet, there are twice as many men as women.
The organiser takes £600 altogether.
There are 50 men at the buffet.
How much does each person pay for the buffet?

FM AU 12 The cost of show tickets for adults and children are in the ratio 5 : 3.
30 adults and 40 children visit the show.
Children's tickets cost £3.60.
How much money will the show take altogether?

HOMEWORK 2I

1 Peter and Margaret's ages are in the ratio 4 : 5. If Peter is 16 years old, how old is Margaret?

2 Cans of lemonade and packets of crisps were bought for the school disco in the ratio 3 : 2. The organiser bought 120 cans of lemonade. How many packets of crisps did she buy?

FM 3 In his restaurant, Manuel is making fruit punch, a drink made from fruit juice and iced soda water, mixed in the ratio 2 : 3. Manuel uses 10 litres of fruit juice.
   a   How many litres of soda water does he use?
   b   How many litres of fruit punch does he make?

4 Cupro-nickel coins are minted by mixing copper and nickel in the ratio 4 : 1.
   a   How much copper is needed to mix with 20 kg of nickel?
   b   How much nickel is needed to mix with 20 kg of copper?

5 The ratio of male to female spectators at a school inter-form football match is 2 : 1. If 60 boys watched the game, how many spectators were there in total?

6 Marmalade is made from sugar and oranges in the ratio 3 : 5. A jar of 'Savilles' marmalade contains 120 g of sugar.

   a How many grams of oranges are in the jar?

   b How many grams of marmalade are in the jar?

AU 7 Each year Abbey School holds a sponsored walk for charity. The money raised is shared between a local charity and a national charity in the ratio 1 : 2. Last year the school gave £2000 to the local charity.

   a How much did the school give to the national charity?

   b How much did the school raise in total?

PS 8 Fred's blackcurrant juice is made from 4 parts blackcurrant and 1 part water.
Jodie's blackcurrant juice is made from blackcurrant and water in the ratio 7 : 2.
Which juice contains the greater proportion of blackcurrant?
Show how you work out your answer.

9 Sand and cement is mixed in the ratio 3 : 1.
Cement is sold in 25 kg bags.
Sand is sold in 875 kg sacks.
How many sacks of sand would be needed to mix with 20 bags of cement?

AU 10 The ratio of tins of white paint to coloured paint in a shop storeroom is 2 : 5.
There is enough room on the shelves for 60 tins of paint.
How many tins of white paint can be put on the shelf if the ratio of the tins of white paint to coloured paint is also 2 : 5?

2.7 Best buys

HOMEWORK 2J

FM 1 Compare the prices of the following pairs of products and state which, if any, is the better buy.

   a Mouthwash: £1.99 for a twin pack, or £1.49 each with a 3 for 2 offer.

   b Dusters: 79p for a pack of 6 with a 'buy one pack and get one pack free' offer, or £1.20 for a pack of 20.

2 Compare the following pairs of products and state which is the better buy and why.

   a Tomato ketchup: a medium bottle which is 200 g for 55p or a large bottle which is 350 g for 87p.

   b Milk chocolate: a 125 g bar at 77p or a 200 g bar at 92p.

   c Coffee: a 750 g tin at £11.95 or a 500 g tin at £7.85.

   d Honey: a large jar which is 900 g for £2.35 or a small jar which is 225 g for 65p.

3 Boxes of 'Wetherels' teabags are sold in three different sizes.

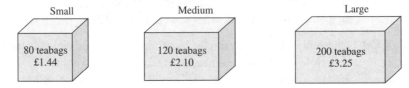

Which size box of teabags gives the best value for money?

4 Bottles of 'Cola' are sold in different sizes. Copy and complete the table.

Size of bottle	Price	Cost per litre
$\frac{1}{2}$ litre	36p	
$1\frac{1}{2}$ litres	99p	
2 litres	£1.40	
3 litres	£1.95	

Which size of bottle gives the best value for money?

AU 5 The following 'special offers' were being promoted by a supermarket.

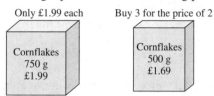

Only £1.99 each   Buy 3 for the price of 2

Cornflakes 750 g £1.99   Cornflakes 500 g £1.69

Which offer is the better value for money? Explain why.

PS 6 Hannah scored 17 out of 20 in a test. John scored 40 out of 50 in a test of the same standard.
Who got the better mark?

2.8 Speed, time and distance

 HOMEWORK 2K

1 A cyclist travels a distance of 60 miles in 4 hours. What was her average speed?

2 How far along a motorway will you travel if you drive at an average speed of 60 mph for 3 hours?

3 Mr Baylis drives on a business trip from Manchester to London in $4\frac{1}{2}$ hours. The distance he travels is 207 miles. What is his average speed?

4 The distance from Leeds to Birmingham is 125 miles. The train I catch travels at an average speed of 50 mph. If I catch the 11.30 am train from Leeds, at what time would I expect to be in Birmingham?

5 Copy and complete the following table.

	Distance travelled	Time taken	Average speed
a	240 miles	8 hours	
b	150 km	3 hours	
c		4 hours	5 mph
d		$2\frac{1}{2}$ hours	20 km/h
e	1300 miles		400 mph
f	90 km		25 km/h

 6 A coach travels at an average speed of 60 km/h for 2 hours on a motorway and then slows down in a town centre to do the last 30 minutes of a journey at an average speed of 20 km/h.
a What is the total distance of this journey?
b What is the average speed of the coach over the whole journey?

7 Hilary cycles to work each day. She cycles the first 5 miles at an average speed of 15 mph and then cycles the last mile in 10 minutes.
a How long does it take Hilary to get to work?
b What is her average speed for the whole journey?

8 Martha drives home from work in 1 hour 15 minutes. She drives home at an average speed of 36 mph.
a Change 1 hour 15 minutes to decimal time in hours.
b How far is it from Martha's work to her home?

PS 9 A tram route takes 15 minutes at an average speed of 16 mph.
The same journey by car is two miles longer.
How fast would a car need to travel to arrive in the same time?

FM 10 A taxi travelled for 30 minutes.
The fare was £24.
If the fare was charged at £1.20 per mile, what was the average speed of the taxi?

AU 11 Two cars are 30 miles apart but travelling towards each other.
The average speed of one car is twice as fast as the other car.
The slower car is averaging 20 mph.
How long is it before they meet up?

Functional Maths Activity

Value Added Tax

You are a shopkeeper and need to use a table to work out the VAT you charge on items for sale.

VAT table

Rate	£1	£10	£100	£500	£1000	£2000	£5000	£10 000
5%	5p	50p	£5					
8%				£40				
15%					£150			
17.5%						£350		
20%							£1000	

Total cost table

Rate	£1	£10	£100	£500	£1000	£2000	£5000	£10 000
5%	£1.05	£10.50	£105					
8%				£540				
15%					£1150			
17.5%						£350		
20%							£6000	

Task 1
Complete the tables.

Functional Maths Activity – continued

Task 2

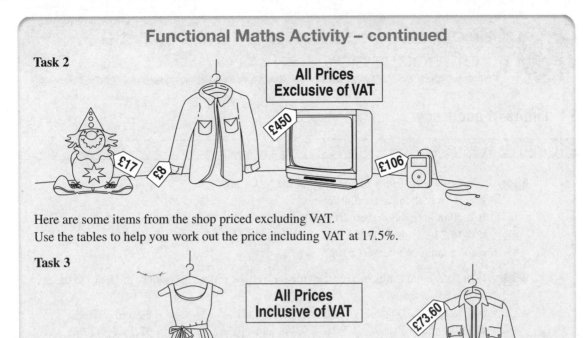

Here are some items from the shop priced excluding VAT.

Use the tables to help you work out the price including VAT at 17.5%.

Task 3

Here are some items priced including VAT.

Use the tables to help you work out the price excluding VAT at 17.5%.

Work out the VAT charged.

Number: Number and limits of accuracy

3.1 Limits of accuracy

HOMEWORK 3A

1 Write down the limits of accuracy of the following.
- **a** 5 cm to the nearest centimetre
- **b** 40 mph to the nearest 10 mph
- **c** 15.2 kg to the nearest tenth of a kilogram
- **d** 75 km to the nearest 5 km

2 Write down the limits of accuracy for each of the following values which are to the given degree of accuracy.

a 7 cm (1 sf)	**b** 18 kg (2 sf)	**c** 30 min (2 sf)	
d 747 km (3 sf)	**e** 9.8 m (1 dp)	**f** 32.1 kg (1 dp)	
g 3.0 h (1 dp)	**h** 90 g (2 sf)	**i** 4.20 mm (2 dp)	
j 2.00 kg (2 dp)	**k** 34.57 min (2 dp)	**l** 100 m (2 sf)	

3 Round off these numbers to the degree of accuracy given.
- **a** 45.678 to 3 sf
- **b** 19.96 to 2 sf
- **c** 0.3213 to 2 dp

4 Write down the upper and lower bounds of each of these values given to the accuracy stated.

a 6 m (1 sf)	**b** 34 kg (2 sf)	**c** 56 min (2 sf)	**d** 80 g (2 sf)
e 3.70 m (2 dp)	**f** 0.9 kg (1 dp)	**g** 0.08 s (2 dp)	**h** 900 g (2 sf)
i 0.70 m (2 dp)	**j** 360 d (3 sf)	**k** 17 weeks (2 sf)	**l** 200 g (2 sf)

PS 5 A theatre has 365 seats.

For a show, 280 tickets are sold in advance.

The theatre's manager estimates that another 100 people to the nearest 10 will turn up without tickets.

Is it possible they will all get a seat, assuming that 5% of those with tickets do not turn up?

Show clearly how you decide.

AU 6 A parking space is 4.8 metres long to the nearest tenth of a metre.

A car is 4.5 metres long to the nearest half a metre.

Which of the following statements is definitely true?

A: The space is big enough

B: The space is not big enough

C: It is impossible to tell whether or not the space is big enough

Explain how you decide.

AU 7 1 litre = 100 cl

A carton contains 1 litre of milk to the nearest 10 cl.

What is the least amount of milk likely to be in the carton?

Give your answer in centilitres.

FM Functional Maths **AU** (AO2) Assessing Understanding **PS** (AO3) Problem Solving

8 Billy has 20 identical bricks. Each brick is 15 cm long measured to the nearest centimetre.

   a What is the greatest length of one brick?

   b What is the smallest length of one brick?

   c If the bricks are put end to end, what is the greatest possible length of all the bricks?

   d If the bricks are put end to end, what is the least possible length of all the bricks?

3.2 Problems involving limits of accuracy

HOMEWORK 3B

1 Cans have a mass of 250 grams to the nearest 10 grams.

   What is the minimum and maximum mass of 10 of these cans?

FM 2 The cans in Question **1** are stacked on a shelf.

   It can safely hold 15 kg of cans.

   What is the maximum number of cans that can safely be put on the shelf?

PS 3 A crate of the cans in Question **1** has a mass of 24 kg.

   How many cans could be in the crate?

4 For each of these rectangles, find the limits of accuracy of the area. The accuracy of each measurement is given.

   a 3 cm × 8 cm   (nearest cm)   **b** 3.2 cm × 6.4 cm   (1 dp)

   c 7.86 cm × 18.78 cm   (2 dp)

5 A rectangular garden has sides of 8 m and 5 m, measured to the nearest metre.

   a Write down the limits of accuracy for each length.

   b What is the maximum area of the garden?

   c What is the minimum perimeter of the garden?

6 A playground is measured as 32 m by 45 m, to the nearest metre. Calculate the limits of accuracy for the area of the playground.

7 The measurements of a box are given as 12 cm × 8 cm × 5 cm to the nearest centimetre. Calculate the limits of accuracy for the volume of the box.

AU 8 Mr Leake is a plumber.

   He has a 10-metre piece of pipe correct to the nearest metre.

   He uses 2 metres of pipe on his first job, 3 metres on his second job and 4 metres on his third job. Each measurement is to the nearest half a metre.

   What is the longest length of pipe he could have left?

AU 9 Belinda is doing a five-mile walk.

   She is walking at an average speed of 3 mph to the nearest whole number of miles per hour.

   She sets off at 2 pm.

   What is the latest time that she will complete the walk?

10 The area of a field is given as 400 m², to the nearest 10 m². One length is given as 24 m, to the nearest metre. Find the limits of accuracy for the other length of the field.

11 In triangle ABC, AB = 8 cm, BC = 6 cm, and ABC = 42°. All the measurements are given to the nearest unit.

   Calculate the limits of accuracy for the area of the triangle.

A*

12 A stopwatch records the time for the winner of a 100-metre race as 12.3 seconds, measured to the nearest one-tenth of a second.

   a What are the greatest and least possible times for the winner?

   b The length of the 100-metre track is correct to the nearest 1 centimetre. What are the greatest and least possible lengths of the track?

   c What is the fastest possible average speed of the winner?

13 A cube has a volume of 27 cm³, to the nearest cm³. Find the range of possible values of the side length of the cube.

14 A cube has a volume of 125 cm³, to the nearest 1 cm³. Find the limits of accuracy of the area of one side of the square base.

PS 15 In the triangle ABC, the length of side AB is 42 cm to the nearest centimetre. The length of side AC is 35 cm to the nearest centimetre. The angle C is 61° to the nearest degree. What is the largest possible size that angle B could be?

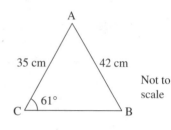

Not to scale

HINTS AND TIPS

You will need to use the Sine Rule.

Functional Maths Activity

Lorries and loads

You are the manager of a haulage company. You own a lorry with 6 axles, with a maximum axle weight limit of 10.5 tonnes. You also own a lorry with 5 axles, with a maximum axle weight limit of 11.5 tonnes. This lorry is 10% cheaper to run per trip than the 6-axle lorry.

These are the load restrictions for heavy goods vehicles.

Type of lorry	Load limit
6 axles, max. axle weight limit 10.5 tonnes	44 tonnes
5–6 axles, max. axle weight limit 11.5 tonnes	40 tonnes

You want to deliver pallets of goods.
Each pallet weighs 500 kg to the nearest 50 kg.

Task 1
A company orders 80 pallets.
Which lorry would you use for this job?
Show clearly any decisions you make and how you decide.

Task 2
A company orders 150 pallets.
Which lorry would you use for this job?
Show clearly any decisions you make and how you decide.

Task 3
A company orders 159 pallets.
Which lorry would you use for this job?
Show clearly any decisions you make and how you decide.

Task 4
Devise a planning chart for orders up to 250 pallets to help you decide which lorry to use.

4 Statistics: Data handling

4.1 Averages

1 **a** For each set of data find the mode, the median and the mean.
  i   6, 4, 5, 6, 2, 3, 2, 4, 5, 6, 1
  ii   14, 15, 15, 16, 15, 15, 14, 16, 15, 16, 15
  iii 31, 34, 33, 32, 46, 29, 30, 32, 31, 32, 33

  b For each set of data decide which average is the best one to use and give a reason.

2 A supermarket sells oranges in bags of ten.
  The weights of each orange in a selected bag were as follows:
    134 g, 135 g, 142 g, 153 g, 156 g, 132 g, 135 g, 140 g, 148 g, 155 g
  a Find the mode, the median and the mean for the weight of the oranges.
  b The supermarket wanted to state the average weight on each bag they sold. Which of the three averages would you advise the supermarket to use? Explain why.

3 The weights, in kilograms, of players in a school football team are as follows:
    68, 72, 74, 68, 71, 78, 53, 67, 72, 77, 70
  a Find the median weight of the team.
  b Find the mean weight of the team.
  c Which average is the better one to use? Explain why.

4 Jez is a member of a local quiz team and, in the last eight games, his total points were:
    62, 58, 24, 47, 64, 52, 60, 65
  a Find the median for the number of points he scored over the eight weeks.
  b Find the mean for the number of points he scored over the eight weeks.
  c The team captain wanted to know the average for each member of the team. Which average would Jez use? Give a reason for your answer.

FM 5 Three dancers were hoping to be chosen to represent their school in a competition.
  They had all been involved in previous competitions.
  The table below shows their scores in recent contests.

Kathy	8, 5, 6, 5, 7, 4, 5
Connie	8, 2, 7, 9, 2
Evie	8, 1, 8, 2, 3

  The teachers said they would be chosen by their best average score.
  Which average would each dancer prefer to be chosen by?

PS 6 **a** Find 3 numbers that have **all** the properties below:
  • a range of 3
  • a mean of 3
  b Find 3 numbers that have **all** the properties below:
  • a range of 3
  • a median of 3
  • a mean of 3

AU 7 A class of students were taking a test.
When asked, "What is the average score for the test?" the teacher said 32 but a student said 28.
They were both correct. Explain how this could be.

4.2 Frequency tables

HOMEWORK 4B

1 Find **i** the mode, **ii** the median and **iii** the mean from each frequency table below.

a A survey of the collar sizes of all the male staff in a school gave these results.

Collar size	12	13	14	15	16	17	18
Number of staff	1	3	12	21	22	8	1

b A survey of the number of TVs in pupils' homes gave these results.

Number of TVs	1	2	3	4	5	6	7
Frequency	12	17	30	71	96	74	25

2 A survey of the number of pets in each family of a school gave these results.

Number of pets	0	1	2	3	4	5
Frequency	28	114	108	16	15	8

a Each child at the school is shown in the data. How many children are at the school?
b Calculate the median number of pets in a family.
c How many families have less than the median number of pets?
d Calculate the mean number of pets in a family. Give your answer to 1 dp.

AU 3 Twinkle travelled to Manchester on many days throughout the year.
The table shows how many days she travelled in each week.

Days	0	1	2	3	4	5
Frequency (no. of weeks)	17	2	4	13	15	1

Explain how you would find the median number of days that Twinkle travelled in a week to Manchester.

4 A survey of the number of television sets in each family home in one school year gave these results:

Number of TVs	0	1	2	3	4	5
Frequency	1	5	36	86	72	56

a How many students are in that school year?
b Calculate the mean number of TVs in a home for this school year.
c How many homes have this mean number of TVs (if you round the mean to the nearest whole number)?
d What percentage of homes could consider themselves average from this survey?

PS 5 A coffee stain removed four numbers (in two columns) from the following frequency table of eggs laid by 20 hens one day.

Eggs	0	1	2	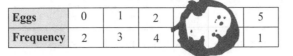	5
Frequency	2	3	4		1

The mean number of eggs laid was 2.5.
What could the missing four numbers be?

4.3 Grouped data

 HOMEWORK 4C

1 For each table of values given below, find:
   i the modal group   **ii** an estimate for the mean.

a

Score	0 – 20	21 – 40	41 – 60	61 – 80	81 – 100
Frequency	9	13	21	34	17

b

Cost (£)	0.00 – 10.00	10.01 – 20.00	20.01 – 30.00	30.01 – 40.00	40.01 – 60.00
Frequency	9	17	27	21	14

FM 2 A hospital has to report the average waiting time for patients in the Accident and Emergency department. A survey was made to see how long casualty patients had to wait before seeing a doctor.
The following table summarises the results for one shift.

Time (minutes)	0 – 10	11 – 20	21 – 30	31 – 40	41 – 50	51 – 60	61 – 70
Frequency	1	12	24	15	13	9	5

   a How many patients were seen by a doctor in the survey of this shift?
   b Estimate the mean waiting time taken per patient.
   c Which average would the hospital use for the average waiting time?
   d What percentage of patients did the doctors see within the hour?

FM 3 A shoe shop undertook a survey to see what size men's shoes were being sold in one month. The following table summarises the results.

Shoe size	3 – 4	5 – 6	7 – 8	9 – 10	11 – 12
Frequency	2	4	21	55	32

   a How many pairs of men's shoes were sold during this month?
   b Estimate the mean men's shoe size sold.
   c Which of the averages is of most use to the shop's manager?
   d What percentage of men's shoes sold were smaller than size 7?

PS 4 The table below shows the total weight of fish caught by the anglers in a fishing competition.

Weight (Kg)	0 < w ≤ 5	5 < w ≤ 10	10 < w ≤ 15	15 < w ≤ 20	20 < w ≤ 25
Frequency	4	15	10	8	3

Helen noticed that two numbers were in the wrong part of the table and that this made a difference of 0.625 to the arithmetic mean.
Which two numbers were the wrong way round?

AU **5** The profit made each week by a tea shop is shown in the table below.

Profit	£0 – £200	£201 – £400	£401 – £600	£601 – £800
Frequency	15	26	8	3

Explain how you would estimate the mean profit made each week.

4.4 Frequency diagrams

HOMEWORK 4D

1 The table shows the time taken by 60 people to travel to work.

Time in minutes	10 or less	Between 10 and 30	30 or more
Frequency	8	19	33

Draw a pie chart to illustrate the data.

2 The table shows the number of GCSE passes that 180 students obtained.

GCSE passes	9 or more	7 or 8	5 or 6	4 or less
Frequency	20	100	50	10

Draw a suitable chart to illustrate the data.

3 Tom is doing a statistics project on the use of computers. He decides to do a survey to find out the main use of computers by 36 of his school friends. His results are shown in the table.

Main use	e-mail	Internet	Word processing	Games
Frequency	5	13	3	15

   a Draw a pie chart to illustrate his data.
   b What conclusions can you draw from his data?
   c Give reasons why Tom's data is not really suitable for his project.

4 In a survey, a TV researcher asks 120 people at a leisure centre to name their favourite type of television programme. The results are shown in the table.

Type of programme	Comedy	Drama	Films	Soaps	Sport
Frequency	18	11	21	26	44

   a Draw a pie chart to illustrate the data.
   b Do you think the sample chosen by the researcher is representative of the population? Give a reason for your answer.

FM 5 Marion is writing an article on health for a magazine. She asked a sample of people the question: 'When planning your diet, do you consider your health?' The pie chart shows the results of her question.

a What percentage of the sample responded 'often'.

b What response was given by about a third of the sample?

c Can you tell how many people there were in the sample? Give a reason for your answer.

d What other questions could Marion ask?

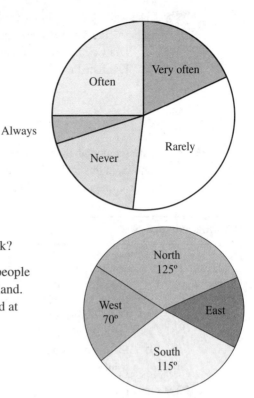

PS 6 A nationwide survey was taken on where people thought the friendliest people were in England. What is the probability that a person picked at random from this survey answered 'East'?

AU 7 You are asked to draw a pie chart representing the different breakfasts that students have in a morning.
What data would you need to obtain in order to do this?

HOMEWORK 4E

1 a The table shows the ages of 300 people at the cinema.

Age, x years	$0 \leq x < 10$	$10 \leq x < 20$	$20 \leq x < 30$	$30 \leq x < 40$	$40 \leq x < 50$
Frequency	25	85	115	45	30

Draw a histogram to show the data.

b At another film show this was the distribution of ages.

Age, x years	$20 \leq x < 30$	$30 \leq x < 40$	$40 \leq x < 50$	$50 \leq x < 60$	$60 \leq x < 70$
Frequency	35	120	130	50	15

Draw a histogram to show this data.

c Comment on the differences between the distributions.

2 The table shows the times taken by 50 children to complete a multiplication square.

Time, s seconds	$10 \leq s < 20$	$20 \leq s < 30$	$30 \leq s < 40$	$40 \leq s < 50$	$50 \leq s < 60$
Frequency	3	9	28	6	4

a Draw a frequency polygon for this data.

b Calculate an estimate of the mean of the data.

3 The waiting times for customers at a supermarket checkout are shown in the table.
  a Draw a histogram of these waiting times.
  b Estimate the mean waiting time.

Waiting time (minutes)	Frequency
$0 \leqslant x < 2$	15
$2 \leqslant x < 4$	7
$4 \leqslant x < 6$	12
$6 \leqslant x < 8$	15
$8 \leqslant x < 10$	12

FM 4 After a mental arithmetic test, all the results were collated for girls and boys separately as shown in this table. The school wants to be able to look at the differences in the results in a clear diagram and use the diagram to estimate the mean scores.

Number correct, N	$0 \leqslant N \leqslant 4$	$5 \leqslant N \leqslant 8$	$9 \leqslant N \leqslant 12$	$13 \leqslant N \leqslant 16$	$17 \leqslant N \leqslant 20$
Boys	5	9	23	28	17
Girls	6	10	19	25	22

  a Draw frequency polygons to illustrate the differences between the boys' scores and the girls' scores.
  b Estimate the mean score for boys and girls scores separately, then comment on your results.

PS 5 The frequency polygon shows the length of time that students spent playing sport in one weekend.
Calculate an estimate of the mean time spent playing sport by the students.

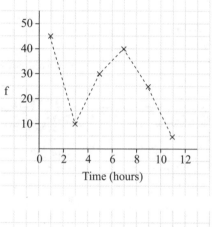

AU 6 The frequency polygon shows the times that a number of people waited at the bus stop before their bus came one morning.
Dan said, 'Most people spent 5 minutes waiting.'
Explain why this is incorrect.

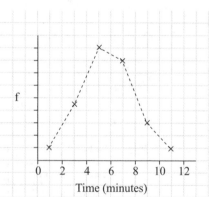

4.5 Histograms with bars of unequal width

HOMEWORK 4F

1 **a** The table shows the ages of 300 people at the cinema.

Age, x years	$0 \leqslant x < 20$	$20 \leqslant x < 30$	$30 \leqslant x < 50$
Frequency	110	115	75

Draw a histogram to show the data.

b Compare your histogram to the one you drew in Homework 3E Question 1a. What do you notice?

Age, x years	$20 \leqslant x < 30$	$30 \leqslant x < 40$	$40 \leqslant x < 50$	$50 \leqslant x < 60$	$60 \leqslant x < 70$
Frequency	35	120	130	50	15

Draw a histogram to show this data.

2 For the histogram on the right.
a write out the frequency diagram
b calculate an estimate of the mean of the distribution.

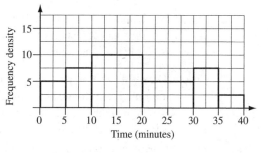

3 The waiting times for customers at a supermarket checkout are shown in the table.
a Draw a histogram of these waiting times.
b Show an estimate of the median on your histogram. Show your working.

Waiting time (minutes)	Frequency
$0 \leqslant x < 1$	15
$1 \leqslant x < 3$	7
$3 \leqslant x < 4$	12
$4 \leqslant x < 5$	15
$5 \leqslant x < 10$	12

AU **4** Andrew was asked to create a histogram.
Explain to Andrew how he can find the height of each bar on the frequency density scale.

A*

FM 5 The histogram shows the science test scores for students in a school.

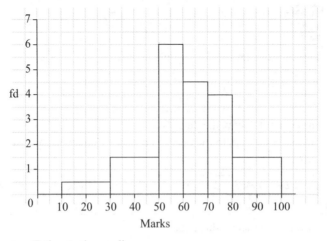

Marks

a   Estimate the median score.

b   Estimate the inter quartile range of the scores.

c   Find an estimate for the mean score.

d   What was the score needed for an A if 10% of the students gained an A?

PS 6 The distances that the members of a church travel to the church are shown in the histogram on the right. It is known that 22 members travel between 4 km and 6 km to church. What is the probability of choosing a member at random who travels more than 4 km to church?

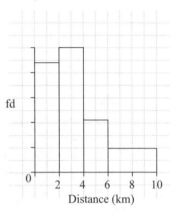

fd

Distance (km)

4.6 Surveys

HOMEWORK 4G

D

1   'People like the DVD rental shop to be open 24 hours a day.'

a   To see whether this statement is true, design a data collection sheet which will allow you to capture data while standing outside a video hire centre.

b   Does it matter at which time you collect your data?

2   The youth club wanted to know which types of activities it should plan, e.g. craft, swimming, squash, walking, disco etc.

a   Design a data collection sheet which you could use to ask the pupils in your school which activities they would want in a youth club.

b   Invent the first 30 entries on the chart.

3   What types of film do your age group watch at the cinema the most? Is it comedy, romance, sci-fi, action, suspense or something else?

a   Design a data collection sheet to be used in a survey of your age group.

b   Invent the first 30 entries on your sheet.

FM 4 Rodrigo, who works for a marketing company, wants to find out who goes to rock concerts. Identifying the target market enables him to make his marketing campaign more successful.

He decides to investigate the hypothesis:

'Boys are less likely to go to rock concerts than girls'.

a Design a data capture form that Rodrigo could use to help him do this.

b Rodrigo records information from a sample of 300 boys and 220 girls. He finds that 190 boys and 160 girls go to rock concerts. Based on this sample, is the hypothesis correct? Explain your answer.

PS 5 What kinds of mobile phones do your classmates own?
Design a data collection sheet to help you find this out.

AU 6 You are asked to find out what kinds of goods the parents of the students at your school like to buy online.
When creating a data collection form, what two things must you include?

4.7 Questionnaires

HOMEWORK 4H

1 Design a questionnaire to test the following statement.
'Young people aged 16 and under will not tell their parents when they have been given detention at school, but the over 16s will always let their parents know.'

2 'Boys will use the Internet almost every day but girls will only use it about once a week.'
Design a questionnaire to test this statement.

3 Design a questionnaire to test the following hypothesis.
'When you are in your twenties, you watch less TV than any other age group.'

4 While on holiday in Wales, I noticed that in the supermarkets there were a lot more women than men, and the only men I did see were over 65.

a Write down a hypothesis from the above observation.

b Design a questionnaire to test your hypothesis.

FM 5 Yohanes and Jacob are doing a survey on the type of DVDs people buy.

a This is one question from Yohanes's survey.

Westerns are only watched by old people.
Don't you agree?
Strongly agree ☐   Agree ☐   Don't know ☐

Give two criticisms of Yohanes's question.

b This is a question from Jacob's survey.

How many DVDs do you buy each month?
3 or fewer ☐   4 or 5 ☐   more than 5 ☐

Give two reasons why this is a good question.

c Make up another good question with responses that could be added to this survey.

PS 6 Design a questionnaire to test the hypothesis,
'Only old people go to Tango lessons'.

AU **7** Kath left a football match and was given a questionnaire containing the following question:
Explain why the response section of this questionnaire is poor.

Question:	How many times do you come to this football ground?			
Response:	Never	☐	More than 5 times	☐
	More than 10 times	☐	More than 20 times	☐

4.8 The data-handling cycle

HOMEWORK 4I

Use the data-handling cycle to describe how you would test each of the following hypotheses,
stating in each case whether you would use Primary or Secondary data.

1 January is the coldest month of the year

2 Girls are better than boys at estimating weights.

3 More men go to cricket matches than women.

4 The TV show *Strictly Come Dancing* is watched by more women than men.

5 The older you are the more likely you are to go ballroom dancing.

4.9 Other uses of statistics

HOMEWORK 4J

FM **1** In 2005, the cost of a litre of diesel was 84p. Using 2005 as a base year, the price index of
diesel for the next five years is shown in the table.

Year	2005	2006	2007	2008	2009	2010
Index	100	104	108	111	113	118
Price	89p					

Work out the price of diesel in each subsequent year. Give your answers to one decimal
place.

PS **2** A country's retail index was given as:
1990: 100
2000: 110
2010: 125
A meal out for four cost £62.50 in the year 2010.
What was the cost of the meal in the year 2000?

AU **3** The Retail Price Index measures how much the daily cost of living increases or decreases.
If 2009 is given a base index number of 100, and 2010 given 102, what does this mean?

4.10 Sampling

HOMEWORK 4K

1 For a school project you have been asked to do a presentation of the timing of the school day. You decide to interview a sample of pupils. How will you choose those you wish to interview if you want your results to be reliable? Give three reasons for your decisions.

2 Comment on the reliability of the following ways of finding a sample.
 a Asking the year 11 Religious Education option class about religion.
 b Find out how many homes have microwaves by asking the first 100 pupils who walk through the school gates.
 c Find the most popular Playstation game by asking a Year 7 form which their favourite is.

3 Comment on the way the following samples have been taken. For those that are not satisfactory, suggest a better way to find a more reliable sample.
 a Bill wanted to find out what proportion of his school went to the cinema, so he obtained an alphabetical list of students and sent a questionnaire to every tenth person on the list.
 b The council wants to know about sports facilities in an area so they sent a survey team to the local shopping centre one Monday morning.
 c A political party wanted to know how much support they had in an area so they rang 500 people from the phone book in the evening.

4 Shameela made a survey of pupils in her school. The size of each year group in the school is shown on the right.

Year	Boys	Girls	Total
7	154	137	291
8	162	156	318
9	134	160	294
10	153	156	309
11	130	140	270
Total	**733**	**749**	**1482**

Claire took a sample of 150 pupils.
 a Explain why this is a suitable size of sample.
 b Draw up a table showing how many pupils of each sex and year she should ask if she wants to obtain a stratified sample.

5 A train company attempted to estimate the number of people who travel by train in a certain town. They telephoned 200 people in the town one evening and asked, 'Have you travelled by train in the last week?' 32 people said 'Yes.' The train company concluded that 16% of the town's population travel by train. Give three criticisms of this method of estimation.

FM 6 Mrs Reynolds, the deputy headteacher at Bradway School, wanted to find out how often the sixth form students in her school went out of the school for their lunch. The number of students in each sixth form year are given in the table.

	Boys	Girls
Y12	88	92
Y13	83	75

 a Create a questionnaire that Mrs Reynolds could use to sample the school.
 b Mrs Reynolds wanted to do a stratified sample using 50 students. How many of each group should she give the questionnaire to?

PS 7 Sandila's school had a total of 1260 students and there were 28 students in her class. One day a survey was carried out using students sampled from the whole school, and four boys and three girls in Sandila's class were chosen to take part in the survey. Estimate how many students in the whole school were involved in the sample.

AU 8 You are asked to conduct a survey at a concert where the attendance is approximately 15 000. Explain how you could create a stratified sample of the crowd.

Functional Maths Activity

Air traffic

The following table shows data about air traffic between the UK and abroad from 1980 to 1990.

	1980	1982	1984	1986	1988	1990
Flights (thousands)	507	511	566	616	735	819
Number of passengers (millions)	43	44	51	52	71	77

1 Draw a suitable diagram to illustrate the changes over this ten-year period.

2 Research similar figures from the Internet for the year 2000 and draw a suitable diagram to show the changes that have occurred over the years 1980, 1990 and 2000.

3 Estimate the number of flights and number of passengers for the year 2010.

Statistics: Statistical representation

5.1 Line graphs

1 The table shows the estimated number of visits to the cinema in Sheffield.

Year	1970	1975	1980	1985	1990	1995	2000	2005
No. of visits (thousands)	280	110	180	330	510	620	750	810

    a Draw a line graph for this data.

    b From your graph estimate the number of visits to the cinema in Sheffield in 1998.

    c In which 5 year period did the number of visits increase the most?

    d Explain the trend in the number of visits. What reasons can you give to explain this trend?

FM 2 Maria started a tea shop and was interested in how trade was picking up over the first few weeks. The table shows the number of teas sold in these weeks.

Week	1	2	3	4	5
Teas sold	67	82	100	114	124

    a Draw a line graph for this data.

    b From your graph, estimate the number of teas Maria hopes to see in week 6.

    c Can you give a reason for the way the numbers of teas increases?

PS 3 A kitten is weighed at the end of each week as:

Week	1	2	3	4	5
Weight (g)	420	480	530	560	580

Estimate how much the kitten would weigh after 8 weeks.

AU 4 When plotting a graph to show the winter midday temperatures in Mexico, Pete decided to start his graph at the temperature 10°C.
Explain why he might have done that.

5.2 Stem-and-leaf diagrams

1 The weights of 15 tomatoes are measured:

        62 g, 58 g, 60 g, 48 g, 55 g
        53 g, 62 g, 67 g, 57 g, 54 g
        60 g, 57 g, 62 g, 47 g, 67 g

    a Show the results in an ordered stem-and-leaf diagram, using this key:
        6|3 represents 63 g

    b What was the greatest weight recorded?

    c What was the most common weight measured?

    d What is the difference between the largest and smallest weights measured?

2 A group of friends compared how many new DVDs they had watched during the Christmas break.

11, 19, 20, 4, 18, 22, 4, 8, 21, 14, 18, 23, 8, 8, 17, 23

a Show the results in an ordered stem-and-leaf diagram, using this key: 1|3 represents 13 new DVDs watched

b What was the highest number of new DVDs that any of the group watched?

c What was the most common number of new DVDs watched in the break?

FM 3 Chris wanted to know how many people attended a church each Sunday for a month. He recorded the data as:

74, 80, 81, 70, 79, 85, 68, 69, 80, 75, 79, 84, 70, 71, 76, 92, 89, 87, 73, 85

a Show these results in an ordered stem-and-leaf diagram.

b What is the median number of people attending the church?

c What is the range of the numbers?

PS 4 The stem-and-leaf diagram below shows some weights of boys (left) and girls (right) in the same form.

```
            5 | 4 | 2 4 4 6 6 8
  1 4 4 5 7 8 8 | 5 | 0 0 4 5 5 6 7 8
      2 3 4 4 6 | 6 | 3
          1 2 2 | 7 |
```

Explain what the diagram is telling you about the students in the form.

AU 5 The numbers of sweets in a set of packets were each counted, with the following results.

25, 26, 26, 27, 26, 26, 25, 26, 27, 28, 27, 26, 26, 24, 24, 27

Explain why a stem-and-leaf diagram is not a good way to represent this information.

5.3 Scatter diagrams

HOMEWORK 5C

1 The table shows the heights and weights of twelve students in a class.

a Plot the data on a scatter diagram.

b Draw the line of best fit.

c Jayne was absent from the class, but she knows she is 132 cm tall. Use the line of best fit to estimate her weight.

d A new girl joined the class who weighed 55 kg. What height would you expect her to be?

Student	Weight (kg)	Height (cm)
Ann	51	123
Bridie	58	125
Ciri	57.5	127
Di	62	128
Emma	59.5	129
Flo	65	129
Gill	65	133
Hanna	65.5	135
Ivy	71	137
Joy	75.5	140
Keri	70	143
Laura	78	145

2 The table shows the marks for ten pupils in their mathematics and music examinations.

Pupil	Maths	Music
Alex	52	50
Ben	42	52
Chris	65	60
Don	60	59
Ellie	77	61
Fan	83	74
Gary	78	64
Hazel	87	68
Irene	29	26
Jez	53	45

a Plot the data on a scatter diagram. Take the x-axis for the mathematics scores and mark it from 20 to 100. Take the y-axis for the music scores and mark it from 20 to 100.

b Draw the line of best fit.

c One of the pupils was ill when they took the music examination. Which pupil was it most likely to have been?

d Another pupil, Kris, was absent for the music examination but scored 45 in mathematics, what mark would you expect him to have got in music?

e Another pupil, Lex, was absent for the mathematics examination but scored 78 in music, what mark would you expect him to have got in mathematics?

FM 3 The table shows the time taken and distance travelled by a delivery van for 10 deliveries in one day.

Dist (km)	8	41	26	33	24	36	20	29	44	27
Time (min)	21	119	77	91	63	105	56	77	112	70

a Draw a scatter diagram with time on the horizontal axis.

b Draw a line of best fit on your diagram.

c A delivery takes 45 minutes. How many kilometres would you expect the journey to have been?

d How long would you expect a journey of 30 kilometres to take?

PS 4 Harry records the time taken, in hours, and the distance travelled, in miles, for several different journeys.

Time (h)	1	1.6	2.2	2.6	3.2	3.5	4	4.8	5.2
Distance (m)	42	62	86	104	130	105	165	190	210

Estimate the distance travelled for a journey lasting 200 minutes.

AU 5 Describe what you would expect the scatter graph to look like if someone said that it showed positive correlation.

5.4 Cumulative frequency diagrams

 HOMEWORK 5D

1 An army squad were all sent on a one mile run. Their coach recorded the times they actually took. This table shows the results.

Time (seconds)	Number of players
$200 < x \leqslant 240$	3
$240 < x \leqslant 260$	7
$260 < x \leqslant 280$	12
$280 < x \leqslant 300$	23
$300 < x \leqslant 320$	7
$320 < x \leqslant 340$	5
$340 < x \leqslant 360$	5

a Copy the table and complete a cumulative frequency column.

b Draw a cumulative frequency diagram.

c Use your diagram to estimate the median time and the interquartile range.

B

2 A company had 360 web pages. They recorded how many times they were visited on one day.

a Copy the table and complete a cumulative frequency column.

b Draw a cumulative frequency diagram.

c Use your diagram to estimate the median use of the web pages and the interquartile range.

d Pages with less than 60 visitors are going to be rewritten. About how many pages would need to be rewritten?

Number of visits	Number of pages
$0 < x \leqslant 50$	6
$50 < x \leqslant 100$	9
$100 < x \leqslant 150$	15
$150 < x \leqslant 200$	25
$200 < x \leqslant 250$	31
$250 < x \leqslant 300$	37
$300 < x \leqslant 350$	32
$350 < x \leqslant 400$	17
$400 < x \leqslant 450$	5

 3 For the mock exams, two classes were given two papers – Paper 1 and Paper 2. The results were summarised in the cumulative frequency graphs opposite.

a What is the median score for each paper?

b What is the interquartile range for each paper?

c Which is the harder paper? Explain how you know.

d The teachers wanted 90% of the students to pass each paper and 15% of the students to get top marks in each paper.

What marks for each paper give:

i a pass

ii the top grade.

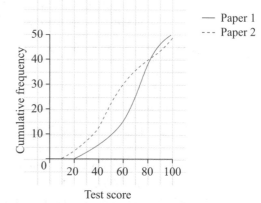

 4 The lengths of time, in minutes, that it took for the Ambulance Service to get an ambulance to a patient and then to a hospital, were recorded. A cumulative frequency diagram of this data is shown opposite.

Calculate the estimated mean length of time it took the Ambulance Service to get to a patient and then to a hospital.

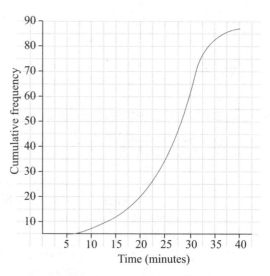

 Johnny was given a cumulative frequency diagram showing the number of students gaining marks in a spelling test. He was told the top 15% were given the top grade. How would you find the marks needed to gain this top award?

5.5 Box plots

HOMEWORK 5E

1 The box plot below shows the number of peas in pods grown by a prize-winning gardener.

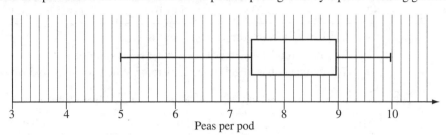

Peas per pod

A young gardener also grew some peas. These are the results for the number of peas per pod. Smallest number 3, Lower quartile 4.75, Median 5.5, Upper quartile 6.25, Highest number 9.

a Copy the diagram and draw a box plot for the young gardener.

b Comment on the differences between the two distributions.

2 The box plot shows the monthly salaries of the men in a computer firm.

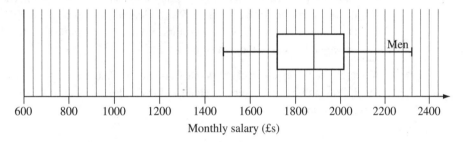

Monthly salary (£s)

The data for the women in the company is Smallest 600, Lower quartile 1300, Median 1600, Upper Quartile 2000, Largest 2400.

a Copy the diagram and draw a box plot for the women's salaries.

b Comment on the differences between the two distributions.

3 The box plots for the hours of life of two brands of batteries are shown below.

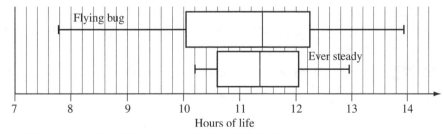

Hours of life

a Comment on the differences in the two distributions.

b Mushtaq wants to get some batteries for his palm top. Which brand would you recommend and why?

4 The following table shows some data on the times of telephone calls to two operators at a mobile phone helpline.

	Lowest time	Lowest Quartile	Median time	Upper Quartile	Highest time
Jack	1 m 10 s	2 m 20 s	3 m 30 s	4 m 50 s	7 m 10 s
Jill	40 s	2 m 20 s	5 m 10 s	7 m 30 s	10 m 45 s

a Draw box plots to compare both sets of data.
b Comment on the differences between the distributions.
c The company has to get rid of 1 operator. Who should go and why?

5 A school entered 80 pupils for an examination. The results are shown in the table.

Mark, x	$0 < x \leqslant 20$	$20 < x \leqslant 40$	$40 < x \leqslant 60$	$60 < x \leqslant 80$	$80 < x \leqslant 100$
Number of pupils	2	14	28	26	10

a Calculate an estimate of the mean.
b Complete a cumulative frequency table and draw a cumulative frequency diagram.
c **i** Use your graph to estimate the median mark.
   ii 12 of these pupils were given a grade A. Use your graph to estimate the lowest mark for which grade A was given.
d Another school also entered 80 pupils for the same examination. Their results were Lowest mark 40, Lower quartile 50, Median 60, Upper quartile 70, Highest mark 80. Draw a box plot to show these results and use it to comment on the differences between the two schools' results.

FM 6 A dental practice had two doctors: Dr Ball and Dr Charlton.
The following box plots were created to illustrate the waiting times for their patients during November.

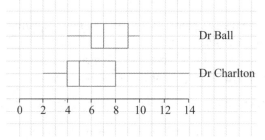

Gabriel was deciding which doctor to try and see. Which one would you advise he see and why?

PS 7 The box plots for a schools end of year science tests are shown below.

What is the difference between the means of the boys' and the girls' test results?

AU 8 Joy was given a diagram showing box plots for the daily sunshine in the seaside resorts of Scarborough and Blackpool for July, but no scale was shown. She was told to write a report on the differences between the sunshine in both resorts.

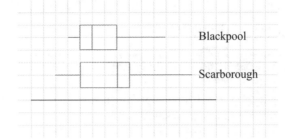

Invent a report that could be possible for her to make from box plots with no scales shown.

Functional Maths Activity

Buying wine in the UK

This table shows the number of cases of different types of wine bought in the UK since 2005.

Wines	Number of cases (millions)				
	2005	2006	2007	2008	2009
Chardonnay	6	8	8	8	9
Pinot Grigio	2	3	4	5	5
Sauvignon Blanc	3	4	5	5	6
Merlot	4	4	5	4	5
Shiraz	3	3	4	5	5
Cabernet Sauvignon	4	5	4	5	5
Rioja	2	2	2	2	2

Use appropriate statistical diagrams and measures to summarise the data given in the table.
Then write a report about the sales of wines in the UK over these five years.

6 Probability: Probabilities of events

6.1 Experimental probability

1 Kylie takes a ball at random from a bag that contains six red and four white balls. The table shows her results.

Number of draws	10	20	50	100	500
Number of white balls	2	6	18	42	192

   a Calculate the relative frequency of a white ball at each stage that Kylie recorded her results.

   b What is the theoretical probability of taking a white ball from the bag?

   c If Kylie took a ball out of the bag a total of 5000 times, how many white balls would you expect her to get?

2 Jason made a six-sided spinner. To test it he threw it 600 times. The table shows the results.

Number	1	2	3	4	5	6
Total	98	152	85	102	62	101

   a Work out the relative frequency of each number.

   b How many times would you expect each number to occur if the spinner is fair?

   c Do you think that the spinner is fair? Give a reason for your answer.

3 A sampling bottle contains red, white and blue balls. There are 50 balls in the bottle. Evie performs an experiment to see how many balls of each colour are in the bottle. She shakes the sampling bottle and looks at the one ball that can be seen in the plastic top. She keeps a tally of each colour as she sees it and summarises her totals in the following chart after 100, 250, 400 and 500 trials.

   a Calculate the relative frequencies of the balls at each stage to 3 sf.

   b How many of each colour do you think are in the bag? Explain your answer.

Red	White	Blue	Total
31	52	17	100
68	120	62	250
102	203	95	400
127	252	121	500

4 Which of these methods would you use to estimate or state the probability of each of the events **a** to **g**?

   Method A: Equally likely outcomes   Method B: Survey or experiment

   Method C: Look at historical data

   a There will be an earthquake in Japan.

   b The next person to walk through the door will be female.

   c A Premier League team will win the FA Cup.

   d You will win a raffle.

   e The next car to drive down the road will be foreign.

   f You will have a Maths lesson this week.

   g A person picked at random from your school will go abroad for their holiday.

5 A bag contains a number of counters. Each counter is coloured red, blue or yellow. Each counter is numbered 1 or 2. The table shows the probability of the colour and number for those counters.

Number of counter	Colour of counter		
	Red	Blue	Yellow
1	0.2	0.3	0.1
2	0.2	0.1	0.1

a A counter is taken from the bag at random. What is the probability that:

   i it is red **and** numbered 2   **ii** it is blue **or** numbered 2

   iii it is red **or** numbered 2?

b There are two yellow counters in the bag. How many counters are in the bag altogether?

6 A check was done one week to see how many passengers on the Number 79 bus were pensioners:

	Mon	Tue	Wed	Thu	Fri
Passengers	950	730	1255	796	980
Pensioners	138	121	168	112	143

For each day, what is the probability that a pensioner was the 400th passenger to board the bus that day?

PS 7 Andrew made a six-sided spinner.

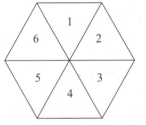

He tested it to see if it was fair.

He spun the spinner 240 times and recorded the results in a table.

Number spinner lands on	1	2	3	4	5	6
Frequency	43	38	31	41	42	44

Do you think the spinner is fair?

Give reasons for your answer.

AU 8 Aleena tossed a coin 50 times to see how many tails she would get.

She said, 'If this is a fair coin, then I should get 25 tails.'

Explain why she is wrong.

6.2 Mutually exclusive and exhaustive events

HOMEWORK 6B

1 Say whether these pairs of events are mutually exclusive or not.
  a Tossing two heads with two coins/tossing two tails with two coins.
  b Throwing an even number with a dice/throwing an odd number with a dice.
  c Drawing a Queen from a pack of cards/drawing an Ace from a pack of cards.
  d Drawing a Queen from a pack of cards/drawing a red card from a pack of cards.
  e Drawing a red card from a pack of cards/drawing a Heart from a pack of cards.

2 Which of the pairs of mutually exclusive events in Question **1** are also exhaustive?

3 A letter is to be chosen at random from this set of letter-cards.

| M | I | S | S | I | S | S | I | P | P | I |

  a What is the probability that the letter chosen is:
    i an S **ii** a P **iii** a vowel?
  b Which of these pairs of events are mutually exclusive?
    i Picking an S / picking a P.   **ii** Picking an S / picking an I.
    iii Picking an I / picking a consonant.
  c Which pair of mutually exclusive events in part **b** is also exhaustive?

FM 4 Two people are to be chosen for a job from these six people.
    Ann   Joan   Jack   John   Arthur   Ethel
  a List all of the possible pairs (there are 15 altogether).
  b What is the probability that the pair of people chosen will:
    i both be female **ii** both be male **iii** both have the same initial
    iv have different initials?
  c Which of these pairs of events are mutually exclusive?
    i Picking two women / picking two men.
    ii Picking two people of the same sex / picking two people of opposite sex.
    iii Picking two people with the same initial / picking two men.
    iv Picking two people with the same initial / picking two women.
  d Which pair of mutually exclusive events in part **c** is also exhaustive?

5 For breakfast I like to have toast, porridge or cereal. The probability that I have toast is $\frac{1}{3}$, the probability that I have porridge is $\frac{1}{2}$. What is the probability that I have cereal?

6 A person is chosen at random. Here is a list of events.
      Event A: the person chosen is male   Event B: the person chosen is female
      Event C: the person chosen is over 18   Event D: the person chosen is under 16
      Event E: the person chosen has a degree   Event F: the person chosen is a teacher
  For each of the pairs of events **i** to **x**, say whether they are:
  a mutually exclusive   **b** exhaustive
  If they are not mutually exclusive, explain why.
    i Event A and Event B   **ii** Event A and Event C
    iii Event B and Event D   **iv** Event C and Event D
    v Event D and Event F   **vi** Event E and Event F
    vii Event E and Event D   **viii** Event A and Event E
    ix Event C and Event F   **x** Event C and Event E

7 An amateur weather man, Steve, records the weather over a year in his village. He knows that the probability of a windy day is 0.4 and that the probability of a rainy day is 0.6. Steve says 'This means it will be either rainy or windy each day as 0.4 + 0.6 = 1, which is certain.' Explain why Steve is wrong.

PS 8 Four brothers, David, Malcolm, Brian and Kevin, regularly run races against each other in the park.
The chances of:

David winning the race is 0.3.

Malcolm winning the race is $\frac{1}{5}$.

Brian winning the race is 45%.

What is the chance of Kevin winning the race?

AU 9 Gareth will either walk, go by bus or be given a lift by his dad to get to football training.
If he walks, the probability that he is late for the practice is 0.4.
If he goes by bus, the probability that he is late for the practice is 0.5.
Explain why it is not necessarily true that, if his dad gives him a lift, his probability of being late for school is 0.1.

6.3 Expectation

HOMEWORK 6C

1 I throw an ordinary dice 600 times. How many times can I expect to get a score of 1?

2 I toss a coin 500 times. How many times can I expect to get a tail?

3 I draw a card from a pack of cards and replace it. I do this 104 times. How many times would I expect to get:
a a red card **b** a Queen **c** a red seven
d the Jack of Diamonds?

4 The ball in a roulette wheel can land on any number from 0 to 36. I always bet on the same block of numbers 1–6. If I play all evening and there is a total of 111 spins of the wheel in that time, how many times could I expect to win?

5 I have five tickets for a raffle and I know that the probability of my winning the prize is 0.003. How many tickets were sold altogether in the raffle?

6 In a bag there are 20 balls, ten of which are red, three yellow, and seven blue. A ball is taken out at random and then replaced. This is repeated 200 times. How many times would I expect to get:
a a red ball **b** a yellow or blue ball
c a ball that is not blue **d** a green ball?

7 A sampling bottle contains black and white balls. It is known that the probability of getting a black ball is 0.4. How many white balls would you expect to get in 200 samples if one ball is sampled each time?

8 **a** Fred is about to take his driving test. The chance he passes is $\frac{1}{3}$.
His sister says 'Don't worry if you fail because you are sure to pass within three attempts because $3 \times \frac{1}{3} = 1$'. Explain why his sister is wrong.
b If Fred does fail would you expect the chance that he passes next time to increase or decrease? Explain your answer.

FM 9 Kara rolls two dice 200 times.
a How many times would she expect to roll a double?
b How many times would she expect to roll a total score greater than 7?

10 An opinion poll used a sample of 200 voters in one area. 112 said they would vote for Party A. There are a total of 50 000 voters in the area.
   a If they all voted, how many would you expect to vote for Party A?
   b The poll is accurate within 10%. Can Party A be confident of winning?

PS 11 A roulette wheel has 37 spaces for the ball to land on. The spaces are numbered 0 to 36. I always bet on a prime number.
   If I play the game all evening and anticipate playing 100 times, how many times would I expect to win on the roulette table?

AU 12 A headteacher is told that the probability of any student being left-handed is 0.14.
   How will she find out how many of her students she should expect to be left-handed?

6.4 Two-way tables

 HOMEWORK 6D

1 The two-way table shows the number of children and the number of computers in 40 homes in one road in a town.

	Number of children			
Number of computers	0	1	2	3
0	3	0	0	0
1	4	10	2	2
2	0	4	9	3
3	0	0	2	1

   a How many homes have exactly two children and two computers?
   b How many homes altogether have two computers?
   c What percentage of the homes have two computers?
   d What percentage of the homes with just one child have one computer?

FM 2 The two-way table shows the part-time earnings of a set of students during one summer break.

Earnings per week	Male	Female
£0 < E ≤ £50	4	1
£50 < E ≤ £100	4	1
£100 < E ≤ £150	11	4
£150 < E ≤ £200	24	14
£200 < E ≤ £250	16	10
More than £250	2	1

   a What percentage of the male students earned between £100 and £150 per week?
   b What percentage of the female students earned between £100 and £150 per week?
   c Estimate the mean earnings of this set of students.
   d Which sex has the greater estimated mean earnings? Explain how you might do this without doing the actual calculation.

3   Here are two fair spinners.

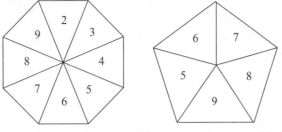

The spinners are spun and the two numbers are added together.

a   Draw a probability sample space diagram.

b   What is the most unlikely score?

c   What is the probability of getting a total of 10?

d   What is the probability of getting a total of 9 or more?

e   What is the probability of getting a total that is an even number?

PS 4   Two six-sided spinners are spun.

Spinner X has the numbers 2, 4, 6, 8, 9 and 11.

Spinner Y has the numbers 3, 4, 5, 6, 7, and 8.

What is the probability that when the two spinners are spun, the two numbers given will multiply to a total greater than 30?

AU 5   Connie planted some tomato plants and kept them in the kitchen, while her husband Harold planted some in the garden.

After the summer, they compared their tomatoes.

	Connie	Harold
Mean diameter	1.9 cm	4.3 cm
Mean number of tomatoes per plant	23.2	12.3

Use the data in the table to explain who had the better crop of tomatoes.

6.5 Addition rule for events

HOMEWORK 6E

1   Shaheeb throws an ordinary dice. What is the probability that he throws:

a   an even number   **b**   5   **c**   an even number or 5?

2   Jane draws a card from a pack of cards. What is the probability that she draws:

a   a red card   **b**   a black card   **c**   a red or a black card?

3   Natalie draws a card from a pack of cards. What is the probability that she draws one of the following:

a   Ace   **b**   King   **c**   Ace or King?

4   A letter is chosen at random from the letters in the word STATISTICS. What is the probability that the letter will be:

a   S   **b**   a vowel   **c**   S or a vowel?

5   A bag contains 10 white balls, 12 black balls and eight red balls. A ball is drawn at random from the bag. What is the probability that it will be:

a   white   **b**   black   **c**   black or white

d   not red   **e**   not red or black?

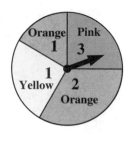

6 A spinner has numbers and colours on it, as shown in the diagram. Their probabilities are given in the tables.

Orange	0.5
Yellow	0.25
Pink	0.25

1	0.4
2	0.35
3	0.25

When the spinner is spun what is the probability of each of the following?

a Orange or pink **b** 2 or 3 **c** 3 or pink **d** 2 or yellow

e i Explain why the answer to **c** is 0.25 and not 0.5.

ii What is the answer to P(2 or orange)?

7 Debbie has 10 CDs in her multi-changer, four of which are rock, two are dance and four are classical. She puts the player on random play. What is the probability that the first CD will be:

a rock or dance **b** rock or classical **c** not rock?

8 Frank buys one dozen free-range eggs. The farmer tells him that a quarter of the eggs his hens lay have double yolks.

a How many eggs with double yolks can Frank expect to get?

b He cooks three and finds they all have a single yolk. He argues that he now has a 1 in 3 chance of a double yolk from the remaining eggs. Explain why he is wrong.

9 John has a bag containing six red, five blue and four green balls. One ball is picked from the bag at random. What is the probability that the ball is:

a red or blue **b** not blue **c** pink **d** red or not blue?

10 Neil and Mandy put music onto their mp3 player so that at their party they have a variety of background music. They put 100 different tracks onto the player as follows: 10 love songs, 15 rap tracks, 35 rock tracks and 40 contemporary tracks.

They set it to play the tracks at random and continuously.

a What is the probability that:

i the first track played is a love song?

ii the last track of the evening is either rock or a contemporary track?

iii the track when they start eating is not a rap track?

b At the party, at midnight, they want to announce their engagement. They want to have a love song or a contemporary track playing. What is the probability that they will not get a track of their choice?

c The party lasts for six hours. For how much time, in hours and minutes, would you expect the mp3 player to have been playing rock tracks?

PS 11 Joy, Vicky and Max play cards together every Sunday night. Joy is always the favourite to win with a probability of 0.65.

In the year 2009 there were 52 Sundays and Vicky won 10 times.

How many times in the year would you expect Max to have won?

AU 12 Kathy has blue, black, brown and yellow jumpers in her bedroom wardrobe.

Brian is asked to bring a jumper down to her.

Why is P(neither blue nor yellow) not equal to P(not blue) + P(not yellow)?

6.6 Combined events

1 Two dice are thrown together. Draw a probability diagram to show the total score.
  a What is the probability of a score that is:
  i 7   **ii** 5 or 8   **iii** bigger than 9   **iv** between 2 and 5
  v odd   **vi** a non-square number?

2 Two dice are thrown. Draw a probability diagram to show the outcomes as pairs of co-ordinates.
  What is the probability that:
  a the score is a 'double'
  b at least one of the dice shows 3
  c the score on one dice is three times the score on the other dice
  d at least one of the dice shows an odd number
  e both dice show a 5
  f at least one of the dice will show a 5
  g exactly one dice shows a 5?

3 Two dice are thrown. The score on one dice is doubled and the score on the other dice is subtracted.
  Complete the probability space diagram.
  For the event described above, what is the probability of a score of:
  a 1
  b a negative number
  c an even number
  d 0 or 1
  e a prime number

	6					6
Score on second dice	5					
	4					
	3	−1				
	2	0				
	1	1	3	5	7	9	11
		1	2	3	4	5	6

Score on first dice

4 When two coins are tossed together, what is the probability of:
  a two heads or two tails   **b** a head and a tail   **c** at least one head?

5 When three coins are tossed together, what is the probability of:
  a three heads or three tails   **b** two tails and one head   **c** at least one head?

6 When a dice and a coin are thrown together, what is the probability of each of the following outcomes?
  a You get a tail on the coin and a 3 on the dice.
  b You get a head on the coin and an odd number on the dice.

7 Max buys two bags of bulbs from his local garden centre. Each bag has four bulbs. Two bulbs are daffodils, one is a tulip and one is a hyacinth. Max takes one bulb from each bag.

Hyacinth				HH
Tulip	DT			
Daffodil				
Daffodil	DD	DD	TD	
	Daffodil	Daffodil	Tulip	Hyacinth

  a There are six possible different pairs of bulbs. List them all.
  b Complete the sample space diagram.
  c What is the probability of getting two daffodil bulbs?
  d Explain why the answer is not $\frac{1}{6}$.

FM 8 Shehab walked into his local supermarket and saw a competition:

> Roll 2 dice.
>
> Get a total of 2
>
> and win a £20 note.
>
> Only 50p a go.

Shehab thought about having a go.

a Draw the sample space for this event.

b What is the probability of winning a £20 note?

c How many goes should he have in order to expect to have won at least once?

d If he had 100 goes, how many times could he expect to have won?

PS 9 I toss four coins. What is the probability that I will get more heads than tails?

AU 10 I roll a dice four times and add the four numbers showing.

Explain the difficulty in drawing a sample space to show all the possible events.

6.7 Tree diagrams

HOMEWORK 6G

1 A dice is thrown twice. Copy and complete the tree diagram on the right to show all the outcomes. Use your tree diagram to work out the probability of:

a getting two sixes

b getting one six

c getting no sixes.

First event	Second event	Outcome	Probability
	$\frac{1}{6}$ → 6	(6, 6)	$\frac{1}{6} \times \frac{1}{6} = \frac{1}{36}$
$\frac{1}{6}$ → 6	$\frac{5}{6}$ → Not 6	(6, Not 6)	$\frac{1}{6} \times \frac{5}{6} = \frac{5}{36}$
Not 6	6		
	Not 6		

2 A bag contains three red and two blue balls. A ball is taken out, replaced, and then another ball is taken out.

a What is the probability that the first ball taken out will be red?

b Copy and complete the tree diagram, showing the possible outcomes.

First event	Second event	Outcome	Probability
	$\frac{3}{5}$ → R	(R, R)	$\frac{3}{5} \times \frac{3}{5} = \frac{9}{25}$
$\frac{3}{5}$ → R	B		
B	R		
	B		

c Using the tree diagram, what is the probability of the following outcomes?

i two red balls   **ii** exactly one red ball   **iii** I get at least one red ball.

3 A card is drawn from a pack of cards. It is replaced, the pack is shuffled and another card is drawn.

   a What is the probability that either card was a Spade?

   b What is the probability that either card was not a Spade?

   c Draw a tree diagram to show the outcomes of two cards being drawn as described.
Use the tree diagram to work out the probability that:

      i both cards will be Spades

      ii at least one of the cards will be a Spade.

4 A bag of sweets contains five chocolates and four toffees. I take two sweets out at random and eat them.

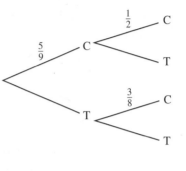

   a What is the probability that the first sweet chosen is:

      i a chocolate   **ii** a toffee?

   b If the first sweet chosen is a chocolate:

      i how many sweets are left to choose from

      ii how many of them are chocolates?

   c If the first sweet chosen is a toffee:

      i how many sweets are left to choose from

      ii how many of them are toffees?

   d Copy and complete the tree diagram.

   e Use the tree diagram to work out the probability that:

      i both sweets will be of the same type   **ii** there is at least one chocolate chosen.

FM 5 Thomas has to take a driving test which is in two parts. The first part is theoretical. He has a 0.4 chance of passing this. The second is practical. He has a 0.5 chance of passing this. Draw a tree diagram covering passing or failing the two parts of the test. What is the probability that he passes both parts?

6 Early every Sunday morning Carol goes out for a run. She has three pairs of shorts of which two are red and one is blue. She has five T-shirts of which three are red and two are blue. Because she doesn't want to disturb her sleeping family she gets dressed in the dark and picks a pair of shorts and a T-shirt at random.

   a What is the probability that the shorts are blue?

   b Copy and complete this tree diagram.

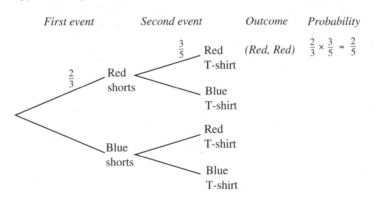

   c What is the probability that Carol goes running in:

      i a matching pair of shorts and T-shirt

      ii a mismatch of shorts and T-shirt

      iii at least one red item?

7 Bob has a bag containing four blue balls, five red balls and one green ball. Sally has a bag containing two blue balls and three red balls. The balls are identical except for the colour. Bob chooses a ball at random from his bag; Sally chooses a ball at random from her bag.

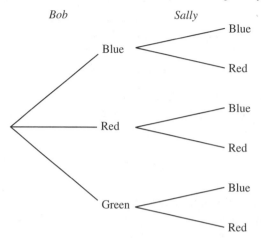

a On a copy of the tree diagram, write the probability of each of the events on the appropriate branch.

b Calculate the probability that both Bob and Sally will choose a blue ball.

c Calculate the probability that the ball chosen by Bob will be a different colour from the ball chosen by Sally.

PS 8 When playing the game 'Rushdown', you are dealt two cards. If you are dealt any two cards which are either 6 or 7 or 8, you have been dealt a 'Tango'.
What is the probability of being dealt a 'Tango'?
Give your answer to 3 decimal places.

AU 9 I have a drawer containing black, blue and brown socks. Explain how a tree diagram can help me find the probability of picking at random two socks of the same colour.

6.8 Independent events 1

 HOMEWORK 6H

1 Ahmed throws a dice twice. The dice is biased so the probability of getting a six is $\frac{1}{4}$.
What is the probability that he gets:
a two sixes   b exactly one six?

2 Betty draws a card from a pack of cards, replaces it, shuffles the pack and then draws another card. What is the probability that the cards are:
a both Hearts   b a Heart and a Spade (in any order)?

3 Colin draws a card from a pack of cards, does not replace it and then draws another card. What is the probability that the cards are:
a both Hearts   b a Heart and a Spade (in any order)?

4 I throw a dice three times. What is the probability of getting a total score of 17 or 18?

5 A bag contains seven white beads and three black beads. I take out a bead at random, replace it and take out another bead. What is the probability that:
a both beads are black   b one bead is black and the other white (in any order)?

6 A bag contains seven white beads and three black beads. I take out a bead at random, do not replace it and take out another bead. What is the probability that:

    a   both beads are black   **b**   one bead is black and the other white (in any order)?

7 When I answer the telephone the call is never for me. Half the calls are for my daughter Janette. One-third of them are for my son Glen. The rest are for my wife Barbara.

    a   I answer the telephone twice this evening. Calculate the probability that:

        i   the first call will be for Barbara   **ii**   both calls will be for Barbara.

    b   The probability that both calls are for Janette is $\frac{1}{4}$. The probability that they are both for Glen is $\frac{1}{9}$. Calculate the probability that they are both for Janette or both for Glen.

8 Anne regularly goes to London by train.

    The probability of the train arriving in London late is 0.08.

    The probability of the train being early is 0.02.

    The probability of it raining in London is 0.3.

    What is the probability of:

    a   Anne getting to London on time and it not raining

    b   Anne travelling to London three days in a row and it raining every day

    c   Anne travelling to London five days in a row and not being late at all?

PS 9 What is the probability of rolling the same number on a dice three times in a row?

AU 10 .Give two events that could be described as 'independent'.

HOMEWORK 6I

1 Steve eats in the school canteen five days a week. The probability that there is turnip on the menu on any day is $\frac{1}{4}$.

    a   What is the probability that in five days there is no turnip on the menu?

    b   What is the probability that Steve gets turnip at least once in five days?

2 Three coins are thrown. What is the probability of getting:

    a   three tails   **b**   at least one head?

3 Adam is a talented all-round athlete. He is entered for the 100 m race, the javelin and the high jump. The probability that he wins each of these events is $\frac{4}{5}$, $\frac{3}{4}$ and $\frac{1}{2}$ respectively.

    a   What is the probability that he doesn't win any event?

    b   What is the probability that he wins at least one event?

4 A bag contains seven red and three blue balls. A ball is taken out and replaced. Another ball is taken out. What is the probability that:

    a   both balls are red   **b**   both balls are blue   **c**   at least one is red?

5 A bag contains seven red and three blue balls. A ball is taken out and not replaced. Another ball is taken out. What is the probability that:

    a   both balls are red   **b**   both balls are blue   **c**   at least one is red?

6   **a**   A coin is thrown three times. What is the probability of:

        i   three heads   **ii**   no heads   **iii**   at least one head?

    b   A coin is thrown four times. What is the probability of:

        i   four heads   **ii**   no heads   **iii**   at least one head?

    c   A coin is thrown five times. What is the probability of:

        i   five heads   **ii**   no heads   **iii**   at least one head?

    d   A coin is thrown n times. What is the probability of:

        i n heads   **ii**   no heads   **iii**   at least one head?

PS 7 The Batemore town show lasts for five days Wednesday to Sunday in a week during April. In this town, at this time of year, the probability of rain on any day during the show is 0.15.

   a What is the probability of:
     i a day with no rain during the show?
     ii two days in a row with no rain?
   b On how many days of the show could they expect:
     i rain?
     ii no rain at all?
   c What is the probability of getting rain on at least one of the days of the show?

PS 8 The probability of planting a banana tree in Bajora and it growing well is 0.7. Kiera plants 10 banana trees in Bajora. What is the probability that at least nine banana trees will grow well?

AU 9 In a class there are 30 students. You are told the probability of a student not being well on the exam day. How will you find the probability of at least one of the students not being well on the exam day?

10 A class has 12 boys and 14 girls. Three of the class are to be chosen at random to do a task for their teacher. What is the probability that:
   a all three are girls   **b** at least one is a boy?

11 From three boys and two girls, two children are to be chosen to present a bouquet of flowers to the Mayoress. The children are to be selected by drawing their names out of a hat. What is the probability that:
   a there will be a boy and a girl chosen
   b there will be two boys chosen
   c there will be at least one girl chosen?

6.9 Independent events 2

HOMEWORK 6J

1 A bag contains two black balls and five red balls. A ball is taken out and replaced. This is repeated three times. What is the probability that:
   a all three are black   **b** exactly two are black
   c exactly one is black   **d** none is black?

2 A bag contains two blue balls and five white balls. A ball is taken out but not replaced. This is repeated three times. What is the probability that:
   a all three are blue   **b** exactly two are blue
   c exactly one is blue   **d** none is blue?

3 A dice is thrown three times. What is the probability that:
   a four sixes are thrown   **b** no sixes are thrown   **c** exactly one six is thrown?

4 Ann is late for school with a probability of 0.7. Bob is late with a probability of 0.6. Cedrick is late with a probability of 0.3. On any particular day what is the probability of:
   a exactly one of them being late   **b** exactly two of them being late?

5 Daisy takes three AS exams in Mathematics. The probability that she will pass Pure 1 is 0.9. The probability that she will pass Statistics 1 is 0.65. The probability she will pass Discrete 1 is 0.95. What is the probability that she will pass:
   a all three modules   **b** exactly two modules   **c** at least two modules?

6 Six out of 10 cars in Britain run on petrol. The rest use diesel fuel. Three cars can be seen approaching in the distance.

a What is the probability that the first one uses petrol?

b What is the probability that exactly two of them use diesel fuel?

c Explain why, if the first car uses petrol, the probability of the second car using petrol is still 6 out of 10.

7 Six T-shirts are hung out at random on a washing line. Three are red and three are blue. Using R and B for red and blue, write down all 20 possible combinations: for example, RRRBBB, RRBBBR and so on. What is the probability of:

a three red shirts being next to each other

b three blue shirts being next to each other?

8 In a class of 30 pupils, 21 have dark hair, seven have fair hair and two have red hair. Two pupils are chosen at random to collect homework.

a What is the probabilty that they:

   i both have fair hair

   ii each have hair of a different colour?

b If three pupils are chosen, what is the probability that exactly two have dark hair?

9 A firm is employing temporary workers. They call 30 for interview. It is found out that seven of the interviewees have a criminal record.

The firm hires three of the interviewees. What is the probability that:

a all three have a criminal record

b only one has a criminal record

c at least two have a criminal record?

AU 10 Paul was playing a card game and was dealt three cards, all Aces. He thought the chance of him now being dealt another Ace was $\frac{1}{52}$.

Explain why he was wrong.

6.10 Conditional probability

HOMEWORK 6K

1 Two tetrahedral (four-sided) dice, whose faces are numbered 1, 2, 3 and 4, are thrown together. The score on the dice is the uppermost face. Show by means of a sample space diagram that there are 16 outcomes with total scores from 2 to 8. What is the probability of:

a a total score of 8   b at least one 3 on either dice?

2 Based on previous results the probability that Manchester United win is $\frac{2}{3}$, the probability they draw is $\frac{1}{4}$ and the probability they lose is $\frac{1}{12}$.

They play three matches. What is the probability that:

a they win all three   b they win exactly two matches

c they win at least one match?

3 In the Spark'n, Spit'n and Fizz'n machine, the probability that the Spark fails is 0.02. The probability that the Spit fails is 0.08 and the probability that the Fizz fails is 0.05.

a What is the probability that nothing fails?

b The machine will still work with one component out of action. What is the probability that it works?

4 On average, Steve is late for school two days each week (of five days).
  a What is the probability that he is late on any one day?
  b In a week of five days, what is the probability that:
    i he is late every day   **ii** he is late exactly once
    iii he is never late   **iv** he is late at least once?

PS 5 The probability of me waking up early at the weekend on exactly one of those days (Saturday or Sunday) is 0.42.
What is the probability of me waking up early on any day?

6 A bag contains five black and three white balls. Three balls are taken out one at a time.
  a If the balls are put back each time, what is the probability of getting:
    i three black balls   **ii** at least one black ball?
  b If the balls are not put back each time, what is the probability of getting:
    i three black balls   **ii** at least one black ball?

7 Two cards are drawn one at a time from a pack of cards. The cards are replaced each time. What is the probability that at least one of them is a Heart?

8 Two cards are drawn from a pack of cards. The cards are not replaced each time. What is the probability that at least one of them is a Heart?

9 A box contains 50 batteries. It is known that 20 of them are dead. John needs two batteries for his calculator. He takes three out. What is the probability that:
  a all three of them are dead   **b** at least one of them works?

10 From a box of 100 batteries, Janet takes out three batteries for her radio. The radio will work if two or three of the batteries are good. What is the probability that the radio will work?

11 Dan has six socks in a drawer, of which four are blue and two are black. He takes out two socks. What is the probability that:
  a both socks are blue   **b** both socks are black
  c he gets a pair of socks   **d** at least one of the socks is blue?

12 One in nine people is left-handed. Five people are in a room. What is the probability that:
  a all five are left-handed   **b** all five are right-handed
  c at least one of them is left-handed?

PS 13 In a pack of cards, the Aces, the Kings, the Queens and the Jacks are all called 'colour cards'.
What is the probability of being dealt four 'colour cards' in a row from a normal deck of cards?

AU 14 A bag of jelly babies contains some yellow, some green and some orange jelly babies, all the same size. Trisha is asked to find the probability of taking out two jelly babies of the same colour.
Explain to Trisha how you would do this, explaining carefully the point where she is most likely to go wrong.

Functional Maths Activity

Lottery competition

A company runs its own version of a lottery but only using the numbers 1 to 20. Four numbers are chosen each week.

1 Andrew chooses the numbers, 1, 2, 3 and 4.
What is his probability of winning the lottery?

2 Evie says Andrew is unlikely to win with four consecutive numbers. She chooses the ages of her nieces and nephews which are 5, 11, 12 and 16.
She says, 'I have a better chance than Andrew.'
Is she correct? Explain your answer.

3 There are 870 employees in the company. Every week they all choose their four numbers and pay 10 pence.
If their four numbers are selected they win £50. There are no rollovers.
In one year, the lottery is run 32 times.
Any profit made goes to a local charity. How much money would the company expect to raise for charity in a year?

7 Number: Fractions and decimals

7.1 Multiplication and division with decimals

HOMEWORK 7A

1 Evaluate each of these.
    a 0.5×0.5 **b** 12.6×0.6 **c** 7.2×0.7
    d 1.4×1.2 **e** 2.6×1.5

2 For each of the following:
    i   estimate the answer by first rounding off each number to the nearest whole number.
    ii calculate the exact answer, and then, by doing a subtraction, calculate how much out your answer to part **i** is.
    a 3.7×2.4 **b** 4.8×3.1 **c** 5.1×4.2
    d 6.5×2.5

PS AU 3 **a** Use any method to work out 15×16
      Use your answer to work out:
    b **i** 1.5×1.6
         ii 0.75×3.2
         iii 4.5×1.6

PS 4 **a** Work out 7.2×3.4
    b Explain how you can use your answer to part **a** to write down the answer to 6.2×3.4

AU 5 **a** Work out 2.3×7.5
    b Given that $4.1 \times 7.5 = 30.75$, use this fact with your answer to part **a** to work out 6.4×7.5

6 Evaluate each of these.
    a 3.12×14 **b** 5.24×15 **c** 1.36×22
    d 7.53×25 **e** 27.1×32

7 Find the total cost of each of the following purchases.
    a Twenty-four litres of petrol at £0.92 per litre.
    b Eighteen pints of milk at £0.32 per pint.
    c Fourteen magazines at £2.25 per copy.

8 A CD case is 0.8 cm thick.
    How many cases are in a pile of CDs that is 16 cm high?

PS 9 **a** Use any method to work out $64 \div 4$
      Use your answer to work out:
    b **i** $6.4 \div 0.04$ **ii** $0.64 \div 4$ **iii** $0.064 \div 0.4$

PS 10 Here are three calculations.
        $43.68 \div 5.6$ $21.7 \div 6.2$ $19.74 \div 2.1$
    Which has the largest answer?
    Show how you know.

7.2 Multiples, factors, prime numbers, powers and roots

HOMEWORK 7B

1 From this list of numbers:

   28   19   36   43   64   53   77   66   56   60   15   29   61   45   51

   Write down those that are:
   a multiples of 4   **b** multiples of 5   **c** prime numbers   **d** factors of 2700

PS 2 During the peak travel time at a railway station, there are trains setting off to the north every 8 minutes, and there are trains to the south every 12 minutes. At 5 pm one train sets off to the north and one sets off to the south. How many more times will two trains be setting off at the same time before 6.30 pm?

3 Write down the negative square root of each of these.
   a 36   **b** 81   **c** 100   **d** 900   **e** 361
   f 169   **g** 225   **h** 1 000 000   **i** 441   **j** 1225

4 Write down the cube root of each of these.
   a 8   **b** 64   **c** 125   **d** 1000   **e** 27 000
   f −27   **g** −1   **h** −216   **i** −8000   **j** −343

AU 5 Here are four numbers.
   8   20   25   64
   Copy and complete this table by putting the numbers in the correct box.

	Square number	Factor of 40
Cube number		
Multiple of 5		

PS 6 Use these four number cards to make a cube number.

$$1 \quad 2 \quad 7 \quad 9$$

AU 7 A number is a factor of 18 and a multiple of 18.
   What is the number?

8 Write down the value of each of these.
   a $\sqrt{0.36}$ **b** $\sqrt{0.81}$ **c** $\sqrt{1.69}$ **d** $\sqrt{0.09}$ **e** $\sqrt{0.01}$
   f $\sqrt{1.44}$ **g** $\sqrt{2.25}$ **h** $\sqrt{1.96}$ **i** $\sqrt{4.41}$ **j** $\sqrt{12.25}$

7.3 Prime factors, LCM and HCF

HOMEWORK 7C

1 Draw factor trees for the following numbers.
   a 144   **b** 75   **c** 98   **d** 420   **e** 560

2 Write the factors for the following in index notation.
   a 48   **b** 54   **c** 216   **d** 1000   **e** 675

AU 3
a Express 36 as a product of prime factors.
b Write your answer to part **a** in index form.
c Use you answer to part **b** to write 18 and 72 as a product of prime factors in index form.

PS 4
$119 = 7 \times 17$
$119^2 = 14\,161$
$119^3 = 1\,685\,159$
a Write 14 161 as a product of prime factors in index form.
b Write 1 685 159 as a product of prime factors in index form.
c Write 119^{10} as a product of prime factors in index form.

PS 5
A mathematician wants to share £18 between three charities so that they each receive a whole number of pounds, and the amount given to each charity is a factor of 18.
How much does he give to each charity?

PS 6
The first three odd prime numbers are all factors of 105.
Explain why this means that seven people can share £105 equally so that each receives an exact number of pounds.

7 Find the LCM of each pair of numbers.
a 5 and 7
b 3 and 8
c 6 and 9
d 10 and 12
e 10 and 15
f 12 and 16
g 16 and 24
h 15 and 35

8 Find the HCF of each pair of numbers.
a 21 and 49
b 27 and 45
c 15 and 25
d 25 and 45
e 48 and 60
f 72 and 108
g 54 and 126
h 99 and 132

9 Write each of the following as a single power of x.
a $x^2 \times x^3$
b $x^4 \times x^5$
c $x^6 \times x$
d $x^5 \times x^5$
e $x^3 \times x^2 \times x^4$

10 Find the HCF of 55 555 and 67 750.

11 Find the LCM of 144 and 162.

FM 12 Nuts are in packs of 12.
Bolts are in packs of 18.
What is the least number of each pack that needs to be bought in order to have the same numbers of nuts and bolts?

PS AU 13 The HCF of two numbers is 5.
The LCM of the same two numbers is 150.
What are the numbers?

7.4 Adding and subtracting fractions

HOMEWORK 7D

1 Evaluate the following.
a $\frac{1}{2} + \frac{1}{5}$
b $\frac{1}{2} + \frac{1}{3}$
c $\frac{1}{3} + \frac{1}{10}$
d $\frac{3}{8} + \frac{1}{3}$
e $\frac{3}{4} + \frac{1}{5}$
f $\frac{1}{3} + \frac{2}{5}$
g $\frac{3}{5} + \frac{3}{8}$
h $\frac{1}{2} + \frac{2}{5}$

2 Evaluate the following.
a $\frac{1}{2} + \frac{1}{4}$
b $\frac{1}{3} + \frac{1}{6}$
c $\frac{3}{5} + \frac{1}{10}$
d $\frac{5}{8} + \frac{1}{4}$

3 Evaluate the following.

a $\frac{7}{8} - \frac{3}{4}$ **b** $\frac{4}{5} - \frac{1}{2}$ **c** $\frac{2}{3} - \frac{1}{5}$ **d** $\frac{3}{4} - \frac{2}{5}$

4 Evaluate the following.

a $\frac{5}{8} + \frac{3}{4}$ **b** $\frac{1}{2} + \frac{3}{5}$ **c** $\frac{5}{6} + \frac{1}{4}$ **d** $\frac{2}{3} + \frac{3}{4}$

AU 5 **a** At a football club, half of the players are English, a quarter are Scottish and one-sixth are Italian. The rest are Irish. What fraction of players at the club are Irish?

b One of the following is the number of players at the club:

30   32   34   36

How many players are at the club?

6 On a firm's coach trip, half the people were employees and two-fifths were partners of the employees. The rest were children. What fraction of the people were children?

7 Five-eighths of a crowd of 35 000 people were male. How many females were in the crowd?

8 What is four-fifths of 65 added to five-sixths of 54?

PS 9 Here are four fractions.

$\frac{1}{6}$ $\frac{5}{12}$ $\frac{1}{4}$ $\frac{1}{3}$

Which three of them add up to 1?

PS FM 10 Pipes are made in $\frac{1}{2}$ m and $\frac{3}{4}$ m lengths.

Show how it is possible to make a pipe exactly 2 m long using the least number of $\frac{1}{2}$ m and $\frac{3}{4}$ m pipes.

Assume that you cannot cut pipes to size.

7.5 Multiplying fractions

HOMEWORK 7E

1 Evaluate the following, leaving your answer in its simplest form.

a $\frac{1}{2} \times \frac{2}{3}$ **b** $\frac{3}{4} \times \frac{2}{5}$ **c** $\frac{3}{5} \times \frac{1}{2}$ **d** $\frac{3}{7} \times \frac{2}{3}$ **e** $\frac{2}{3} \times \frac{5}{6}$

f $\frac{1}{3} \times \frac{3}{5}$ **g** $\frac{2}{3} \times \frac{7}{10}$ **h** $\frac{3}{8} \times \frac{2}{5}$ **i** $\frac{4}{9} \times \frac{3}{8}$ **j** $\frac{4}{5} \times \frac{7}{16}$

2 Kris walked three-quarters of the way along Carterknowle Road which is 3 km long. How far did Kris walk?

AU 3 Jean ate one-fifth of a cake, Les ate a half of what was left. Nick ate the rest. What fraction of the cake did Nick eat?

4 A formula for working out the distance travelled is:

Distance travelled = Speed × Time taken

A snail is moving at one-tenth of a metre per minute. It travels for half a minute. How far has it travelled?

PS 5 George is given £80.

Each week he spends half of the amount left.

What fraction of the £80 will he have left after 4 weeks?

AU 6 You are given that 1 litre = 100 cl.

A bottle holds 1.5 litres.

Tupac drinks half of the contents.

Belinda drinks 25 cl of the contents.

What fraction of the contents are left?

7 Evaluate the following, leaving your answer as a mixed number where possible.

a $1\frac{1}{3} \times 2\frac{1}{4}$ b $1\frac{3}{4} \times 1\frac{1}{3}$ c $2\frac{1}{2} \times \frac{4}{5}$ d $1\frac{2}{3} \times 1\frac{3}{10}$

e $3\frac{1}{4} \times 1\frac{3}{5}$ f $2\frac{2}{3} \times 1\frac{3}{4}$ g $3\frac{1}{2} \times 1\frac{1}{6}$ h $7\frac{1}{2} \times 1\frac{3}{5}$

PS 8 Which is the smaller, $\frac{3}{4}$ of $5\frac{1}{3}$ or $\frac{2}{3}$ of $4\frac{2}{5}$?

AU 9 I estimate that I need 60 litres of lemonade for a party.
I buy 24 bottles, each containing $2\frac{3}{4}$ litres.
Have I bought enough lemonade for the party?

PS 10 Pizzas are often cut into 8 equal pieces.
If a pizza is cut into six equal pieces how much more pizza is in each piece?
Give your answer as a fraction of the whole pizza.

AU 11 A company employs 200 people.
The manager says that exactly two-thirds of the employees are women and three-quarters of the employees are full-time.
One of the statements is true and one is not accurate.
Explain which statement is which.

FM 12 A fish farmer is trying to work out how many fish are in a pond.
He captures 100 fish, marks them and puts them back in the pond.
Later he captures 100 fish and finds that 25 are marked.
Approximately how many fish are in the pond?

7.6 Dividing by a fraction

HOMEWORK 7F

1 Evaluate the following, leaving your answer as a mixed number where possible.

a $\frac{1}{5} \div \frac{1}{3}$ b $\frac{3}{5} \div \frac{3}{8}$ c $\frac{4}{5} \div \frac{2}{3}$ d $\frac{4}{7} \div \frac{8}{9}$

e $4 \div 1\frac{1}{2}$ f $5 \div 3\frac{2}{3}$ g $8 \div 1\frac{3}{4}$ h $6 \div 1\frac{1}{4}$

i $5\frac{1}{2} \div 1\frac{1}{3}$ j $7\frac{1}{2} \div 2\frac{2}{3}$ k $1\frac{1}{2} \div 1\frac{1}{5}$ l $3\frac{1}{5} \div 3\frac{3}{4}$

FM 2 A pet shop has 36 kg of hamster food. Tom, who owns the shop, wants to pack this into bags, each containing three-quarters of a kilogram. How many bags can he make in this way?

AU 3 Bob is putting a fence down the side of his garden, it is to be 20 m long. The fence comes in sections; each one is $1\frac{1}{3}$ m long. How many sections will Bob need to put the fence all the way down the one side of his garden?

PS 4 An African bullfrog can jump a distance of $1\frac{1}{4}$ m in one hop. How many hops would it take an African bullfrog to hop a distance of 100 m?

AU 5 Three-fifths of all 14-year-olds in a school visit the dentist each year.
One-third of those who do not visit the dentist have a problem with their teeth.
What fraction of the 14-year-olds do not visit the dentist and have a problem with their teeth?

FM 6 How many half-litre tins of paint can be poured into a 2.2 litre paint tray without spilling?

I work 8 hours in a day.

Short tasks last $\frac{1}{4}$ hour.

Long tasks last $\frac{3}{4}$ hour.

I have to complete at least three long tasks.
How many short tasks can I also do in one working day?

8 Evaluate the following, leaving your answer as a mixed number wherever possible.

a $\frac{4}{5} \times \frac{1}{2} \times \frac{3}{8}$

b $\frac{3}{4} \times \frac{7}{10} \times \frac{5}{6}$

c $\frac{2}{3} \times \frac{5}{6} \times \frac{9}{10}$

d $1\frac{1}{4} \times \frac{2}{3} \div \frac{5}{6}$

e $\frac{5}{8} \times 1\frac{1}{3} \div 1\frac{1}{10}$

f $2\frac{1}{2} \times 1\frac{1}{3} \div 3\frac{1}{3}$

Functional Maths Activity

Flooring specialist

Imagine you are a flooring specialist.

A customer asks you to quote for laying a new floor in her dining room. The room is 5.3 m long and 4.5 m wide. She would prefer luxury carpet or wood, but would consider other types you recommend.

The prices you charge for various types of flooring are given in the table below, along with other useful information.

Flooring type	Size	Pack size (minimum quantity)	Price	Labour cost to lay the flooring
Carpet tiles	500 mm × 500 mm	Packs of 10	£1.89 per pack	£3.50 per m^2
Plain carpet	4 m wide roll	1 m units	£1.39 per m	£4.00 per m^2
Luxury carpet	4 m wide roll	1 m units	£2.99 per m	£4.00 per m^2
Wood (beech)	1.8 m lengths, 200 mm wide	Packs of 5	£15.30 per pack	£8.50 per m^2
Wood (oak)	1.8 m lengths, 200 mm wide	Packs of 5	£22.50 per pack	£8.50 per m^2
Ceramic tiles	200 mm × 300 mm	Packs of 10	£3.49 per pack	£6.50 per m^2

1 Draw up a costing table so that you can tell the customer how much each flooring type would cost. For each flooring type your table needs to show:

- The amount of flooring material required, bearing in mind the minimum purchase quantities
- The cost of material
- The total labour cost to lay the floor
- The total cost to lay the floor
- The cost per metre to lay the floor, including material cost and labour cost
- The area of material (in m^2) that would be wasted (this should be a minimum)
- The cost of the wasted material
- The fraction of the material cost that is wasted (show this as a percentage).

2 The customer has a budget of £350. Can she afford one of her preferred flooring types?

3 What are the advantages and disadvantages of each flooring type? Consider durability, cost, cleaning and any other properties you can think of.

4 Which would you recommend to the customer and why?

Geometry and measures: Area, volume and angles

8.1 Area of a trapezium

HOMEWORK 8A

D

1 Calculate the perimeter and the area of each of these trapeziums.

a

5 cm
5 cm
4.1 cm
4 cm
9 cm

b

6 cm
7 cm
10 cm
13 cm

2 Calculate the area of each of these shapes.

a

7 m
4 m
3 m
3 m
15 m

b

10 cm
5 cm
3 cm
2 cm
8 cm

3 Which of the following shapes has the largest area?

a

3 cm
2.5 cm
5 cm

b

8 cm
2.4 cm

AU 4 Calculate the area of this trapezium.

5 cm
7.5 cm
6 cm
7.5 cm
14 cm

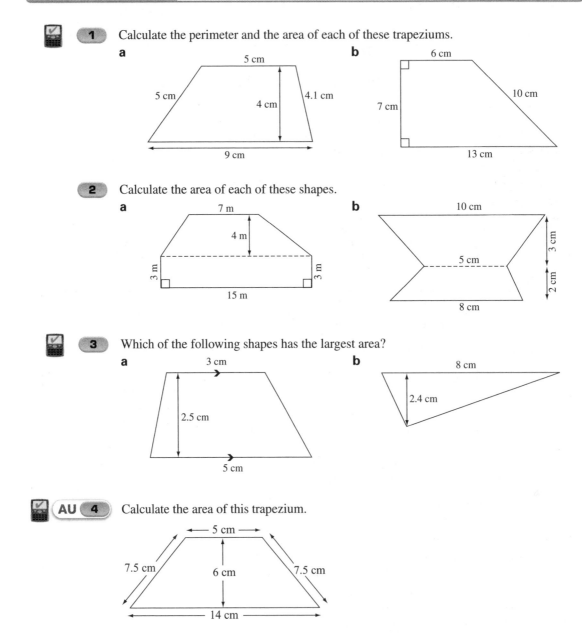

FM Functional Maths **AU** (AO2) Assessing Understanding **PS** (AO3) Problem Solving

FM **5** This is the plan of an area that is to be seeded with grass.

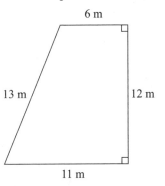

Seed should be planted at a rate of 30 g per m².
How much grass seed will be required?

PS **6** A trapezium has an area of 100 cm². The parallel sides are 17 cm and 23 cm long.
How far apart are the parallel sides?

7 Calculate the area of the shaded part in each of these diagrams.

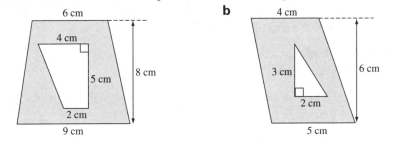

a

b

 8 What percentage of this shape has been shaded?

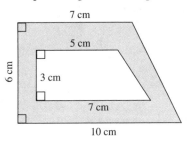

8.2 Sectors

HOMEWORK 8B

 1 For these sectors, calculate the arc length and the sector area.

a

b

50°
10 cm

90°
7 cm

2 Calculate the arc length and the area of a sector whose arc subtends a right angle in a
circle of diameter 10 cm. Give your answer in terms of π.

A

3 Calculate the total perimeter of each of these shapes.

a

20 cm

b

12 cm

4 Calculate the area of each of these shapes.

a

120°

8 cm

b

45°

9 cm

5 There is an infrared sensor in a security system. The sensor can detect movement inside a sector of a circle. The radius of the circle is 16 m. The sector is 120°. Calculate the area of the sector.

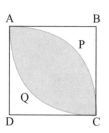

120°

16 m   16 m

Sensor

AU 6 A circle of radius 8 cm is cut up into five congruent sectors. Calculate the perimeter of each one.

FM 7 A shelf to fit in the corner of a room is to be cut in the shape of a quarter of a circle. It will be cut from a square of wood of side 30 cm. What will be the area of the shelf?

A*

PS 8 ABCD is a square of side length 15 cm. APC and AQC are arcs of the circle with centres D and B. Calculate the area of the unshaded part.

A   B
                 P

            Q
D   C

8.3 Volume of a prism

HOMEWORK 8C

C

1 For each prism shown, calculate the area of the cross-section and the volume.

a

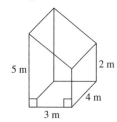

5 m   2 m
        4 m
3 m

b

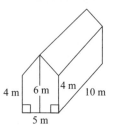

4 m   6 m   4 m   10 m
5 m

AU 2 A chocolate box is in the form of a triangular prism. It is 18 cm long and has a volume of 387 cm³.
What is the area of the triangular end of the box?

FM 3 A wooden door wedge has a cross-section which is this shape:

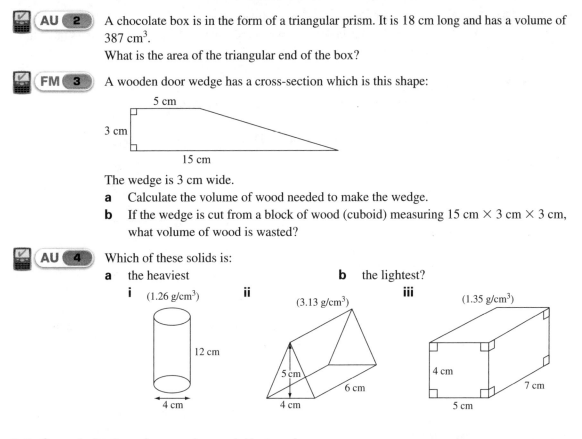

The wedge is 3 cm wide.

a Calculate the volume of wood needed to make the wedge.

b If the wedge is cut from a block of wood (cuboid) measuring 15 cm × 3 cm × 3 cm, what volume of wood is wasted?

AU 4 Which of these solids is:

a the heaviest **b** the lightest?

i (1.26 g/cm³) **ii** (3.13 g/cm³) **iii** (1.35 g/cm³)

8.4 Special triangles and quadrilaterals

HOMEWORK 8D

1 For each of these shapes, calculate the value of the lettered angles.

a 118° b
a 72°

b c d
122° e

c 161° 23°
f g

2 Calculate the values of x and y in each of these shapes.

a $x + 30$ $2y - 10$
$2x$ $3y - 10$

b $11x + 4$
$3y + 5$ $5x$

PS 3 Find the value of x in each of these quadrilaterals with the following angles and state the type of quadrilateral it is.

a $x + 10°, x + 30°, x - 30°, x - 50°$ **b** $x°, x - 10°, 3x - 15°, 3x - 15°$

D

4   **a**   What do the interior angles of a quadrilateral add up to?
   b   Use the fact that the angles of a triangle add up to 180°, to prove that the sum of the interior angles of any quadrilateral is 360°.

FM 5   The diagram shows the side wall of a barn.
The architect says that angle D must not be more than twice as big as angle A.
What is the largest possible size of angle D?

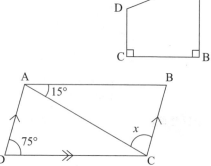

AU 6   The diagram shows a parallelogram ABCD.
AC is a diagonal.
Show that angle x is a right angle.
Give reasons for your answer.

AU 7   Give two reasons to explain why the trapezium is different from the parallelogram.

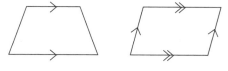

8.5 Angles in polygons

HOMEWORK 8E

1   Calculate the sum of the interior angles of polygons with:
   a   7 sides   **b**   11 sides   **c**   20 sides   **d**   35 sides

2   Calculate the size of the interior angle of regular polygons with:
   a   15 sides   **b**   18 sides   **c**   30 sides   **d**   100 sides

3   Find the number of sides of the polygon with the interior angle sum of:
   a   1440°   **b**   2520°   **c**   6120°   **d**   6840°

4   Find the number of sides of the regular polygon with an exterior angle of:
   a   20°   **b**   30°   **c**   18°   **d**   4°

5   Find the number of sides of the regular polygon with an interior angle of:
   a   135°   **b**   165°   **c**   170°   **d**   156°

PS 6   What is the name of the regular polygon whose interior angle is three times its exterior angle?

7 Anne measured all the interior angles in a polygon. She added them up to make 1325°, but she had missed out one angle. What is the:

    a name of the polygon that Anne measured and   **b** size of the missing angle?

AU 8 This shape is made from a regular pentagon and a regular octagon.

Work out the size of angle *x*.

FM 9 Jamal is cutting metal from a rectangular sheet to make this sign.

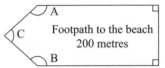

He decides it will look best if angles A and B are the same size and each of them is twice as big as angle C. How big are angles A, B and C?

PS 10 ABCDE is a regular pentagon.

Work out the size of angle ADE.
Give reasons for your answer.

AU 11 Which of the following statements are true for a regular hexagon?

    a The size of each interior angle is 60°
    b The size of each interior angle is 120°
    c The size of each exterior angle is 60°
    d The size of each exterior angle is 240°

Functional Maths Activity

Packaging sweets

A sweet manufacturer wants a new package for an assortment of sweets.

The package must have a volume of 1000 cm³ in order to hold the sweets.

The chosen design will be a prism.

The length has been specified as 20 cm.

The cross-section of the package will be one of three possibilities: a square, an equilateral triangle or a circle.

You have been asked to investigate the amount of packaging material needed for each design, because this will affect the cost of manufacture.

Calculate the surface area of each of the three designs.

Comment on how much difference there is between the three surface areas and how this could affect production costs.

This formula could be useful:

The area of an equilateral triangle of side a is $\frac{1}{4} \sqrt{3}a^2$

9 Number: Powers, standard form and surds

9.1 Powers (indices)

HOMEWORK 9A

1 Write these expressions using power notation. Do not work them out yet.
  a $5 \times 5 \times 5 \times 5$
  b $7 \times 7 \times 7 \times 7 \times 7$
  c $19 \times 19 \times 19$
  d $4 \times 4 \times 4 \times 4 \times 4$
  e $1 \times 1 \times 1 \times 1 \times 1 \times 1 \times 1$
  f $8 \times 8 \times 8 \times 8 \times 8$
  g 6
  h $11 \times 11 \times 11 \times 11 \times 11 \times 11$
  i $0.9 \times 0.9 \times 0.9 \times 0.9$
  j $999 \times 999 \times 999$

2 Write these power terms out in full. Do not work them out yet.
  a 4^5 **b** 8^4 **c** 5^3 **d** 9^6 **e** 1^{11}
  f 7^3 **g** 5.2^3 **h** 7.5^3 **i** 7.7^4 **j** $10\,000^3$

3 Using the power key on your calculator (or another method), work out the values of the power terms in Question **1**.

4 Using the power key on your calculator (or another method), work out the values of the power terms in Question **2**.

FM 5 A box is a cube.
The width of the box is 1 metre.
To work out the volume of a box, use the formula:
Volume = (width)^3
Some dolls are manufactured and packed into the box. The volume of each doll is 0.03 m^3.
Work out the amount of space in the box when 24 dolls are packed into the box.

AU 6 Write each number as a power of a different number.
The first one has been done for you.
  a $27 = 3^3$
  b $16 =$
  c $125 =$
  d $64 =$

7 Without using a calculator, work out the values of these power terms.
  a 7^0 **b** 9^1 **c** 17^0 **d** 1^{91} **e** 10^5

8 Using your calculator, or otherwise, work out the values of these power terms.
  a $(-2)^3$ **b** $(-1)^{11}$ **c** $(-3)^4$ **d** $(-5)^3$ **e** $(-10)^6$

9 Without using a calculator, write down the answers to these.
  a $(-4)^2$ **b** $(-5)^3$ **c** $(-3)^4$ **d** $(-2)^5$ **e** $(-1)^6$

PS 10 $5^5 = 3125$ $5^6 = 15\,625$ $5^7 = 78\,125$ $5^8 = 390\,625$
Write down the last three digits of each of the following powers of 5.
  a 5^{99}
  b 5^{100}

HOMEWORK 9B

AU 1 You are given that $6^4 = 1296$.
Write down the value of 6^{-4}.

2 Write down each of these in fraction form.
   a 5^{-2} **b** 4^{-1} **c** 10^{-3} **d** 3^{-3} **e** x^{-2} **f** $5t^{-1}$

3 Write down each of these in negative index form.
   a $\dfrac{1}{2^4}$ **b** $\dfrac{1}{7}$ **c** $\dfrac{1}{x^2}$

4 Change each of the following expressions into an index form of the type shown.
   a All of the form 2^n **i** 32 **ii** $\frac{1}{4}$
   b All of the form 10^n **i** $10\,000$ **ii** $\frac{1}{100}$
   c All of the form 5^n **i** 625 **ii** $\frac{1}{125}$

5 Find the value of each of the following, where the letters have the given values.
   a Where $x = 3$ **i** x^2 **ii** $4x^{-1}$
   b Where $t = 5$ **i** t^{-2} **ii** $5t^{-4}$
   c Where $m = 2$ **i** m^{-3} **ii** $4m^{-2}$

6 $a = 3$ and $b = 2$. Calculate the value of
   a $3a^{-1} + 2b^{-2}$, giving your answer as a fraction in its simplest form.
   b $a^{-2} + b^{-3}$, giving your answer as a fraction in its simplest form.

PS 7 a and b are integers.
$2^a + 3^b = 41$
Work out the values of a and b.

AU 8 c and d are integers. c is even.
Decide whether $c^2 + d^3$ is even or odd, or whether it could be either. Give examples to show how you decided.

PS AU 9 Put these in order from smallest to largest:
x^0 x^{-1} x^1
   a when x is greater than 1
   b when x is between 0 and 1
   c when x is between -1 and 0

HOMEWORK 9C

1 Write these as single powers of 7.
   a $7^3 \times 7^2$ **b** $7^3 \times 7^6$ **c** $7^4 \times 7^3$ **d** 7×7^5 **e** $7^5 \times 7^9$ **f** 7×7^7

2 Write these as single powers of 5.
   a $5^6 \div 5^2$ **b** $5^8 \div 5^2$ **c** $5^4 \div 5^3$ **d** $5^5 \div 5^5$ **e** $5^6 \div 5^4$

3 Simplify these and write them as single powers of a.
   a $a^2 \times a$ **b** $a^3 \times a^2$ **c** $a^4 \times a^3$ **d** $a^6 \div a^2$ **e** $a^3 \div a$ **f** $a^5 \div a^4$

AU 4   **a** $a^x \times a^y = a^6$
Write down a possible pair of values of x and y.
b $a^x \div a^y = a^6$
Write down a possible pair of values of x and y.

5   Simplify these expressions.
a $3a^4 \times 5a^2$ **b** $3a^4 \times 7a$ **c** $5a^4 \times 6a^2$ **d** $3a^2 \times 4a^7$
e $5a^4 \times 5a^2 \times 5a^2$

6   Simplify these expressions.
a $8a^5 \div 2a^2$ **b** $12a^7 \div 4a^2$ **c** $25a^6 \div 5a$ **d** $48a^8 \div 6a^{-1}$
e $24a^6 \div 8a^{-2}$ **f** $36a \div 6a^5$

7   Simplify these expressions.
a $3a^3b^2 \times 4a^3b$ **b** $7a^3b^5 \times 2ab^3$ **c** $4a^3b^5 \times 5a^4b^{-1}$
d $12a^3b^5 \div 4ab$ **e** $24a^3b^5 \div 6a^2b^{-3}$

8   **a**   Write down two possible multiplication questions with an answer of $18x^3y^4$.
b   Write down two possible division questions with an answer of $18x^3y^4$.

PS 9 a and b are different prime numbers.
What is the smallest value of a^2b^2?

PS 10   Use the general rule for dividing powers of the same number
$a^x \div a^y = a^{x-y}$
to prove that any number raised to the power -1 is the reciprocal of that number.

HOMEWORK 9D

Evaluate the following.

1 $36^{\frac{1}{2}}$ **2** $144^{\frac{1}{2}}$ **3** $25^{\frac{1}{2}}$ **4** $196^{\frac{1}{2}}$ **5** $8^{\frac{1}{3}}$ **6** $125^{\frac{1}{3}}$

7 $32^{-\frac{1}{5}}$ **8** $144^{-\frac{1}{2}}$ **9** $27^{-\frac{1}{3}}$ **10** $(\frac{25}{81})^{\frac{1}{2}}$ **11** $(\frac{81}{36})^{\frac{1}{2}}$ **12** $(\frac{36}{64})^{\frac{1}{2}}$

13 $(\frac{8}{27})^{\frac{1}{3}}$ **14** $(\frac{16}{625})^{\frac{1}{4}}$ **15** $(\frac{4}{9})^{-\frac{1}{2}}$ **16** $(\frac{16}{25})^{-\frac{1}{2}}$ **17** $(\frac{8}{27})^{-\frac{1}{3}}$

AU 18   Which of these is the odd one out?
$27^{-\frac{1}{3}}$ $25^{-\frac{1}{2}}$ 3^{-1}
Show how you decided.

AU 19   Imagine that you are the teacher.
Write down how you would teach the class that $16^{-\frac{1}{4}}$ is equal to $\frac{1}{2}$.

PS 20 $x^{-\frac{1}{4}} = y^{-\frac{1}{2}}$
Find values for x and y that make this equation work.

HOMEWORK 9E

1   Evaluate the following.
a $16^{\frac{3}{4}}$ **b** $125^{\frac{4}{3}}$ **c** $81^{\frac{3}{4}}$

2   Rewrite the following in index form.
a $\sqrt[4]{t^3}$ **b** $\sqrt[5]{m^2}$

3   Evaluate the following.
a $27^{\frac{2}{3}}$ **b** $8^{\frac{4}{3}}$ **c** $36^{\frac{3}{2}}$ **d** $81^{1.25}$

4 **a** Draw the graph of $y = 2^x$ for $x = -2$ to 5.
  b Use your graph to estimate the value of $2^{\frac{5}{2}}$.

5 Using a trial-and-improvement method, or otherwise, solve these equations.
  a $6^x = 60$ **b** $10^x = 2$ (You could try to use the **power** key on your calculator.)

6 **a** Evaluate $8^{\frac{1}{3}}$.   **b** Write $16^{-\frac{1}{2}} \times 2^{-3}$ as a power of 2.
  c Given that $32^y = 2$, find the value of y.

PS **7** Which of these is the odd one out?
  $16^{-\frac{3}{4}}$ $64^{-\frac{1}{2}}$ $8^{-\frac{2}{3}}$
  Show how you decided.

AU **8** Imagine that you are the teacher.
  Write down how you would teach the class that $27^{-\frac{2}{3}}$ is equal to $\frac{1}{9}$.

9.2 Standard form

HOMEWORK 9F

1 Evaluate the following.
  a 3.5×100 **b** 2.15×10 **c** 6.74×1000 **d** 4.63×10
  e 30.145×10 **f** 78.56×1000 **g** 6.42×10^2 **h** 0.067×10
  i 0.085×10^3 **j** 0.798×10^5 **k** 0.658×1000 **l** 215.3×10^2
  m 0.889×10^6 **n** 352.147×10^2 **o** 37.2841×10^3 **p** 34.28×10^6

2 Evaluate the following.
  a $4538 \div 100$ **b** $435 \div 10$ **c** $76459 \div 1000$ **d** $643.7 \div 10$
  e $4228.7 \div 100$ **f** $278.4 \div 1000$ **g** $246.5 \div 10^2$ **h** $76.3 \div 10$
  i $76 \div 10^3$ **j** $897 \div 10^5$ **k** $86.5 \div 1000$ **l** $1.5 \div 10^2$
  m $0.8799 \div 10^6$ **n** $23.4 \div 10^2$ **o** $7654 \div 10^3$ **p** $73.2 \div 10^6$

3 Evaluate the following.
  a 400×300 **b** 50×4000 **c** 70×200 **d** 30×700
  e $(30)^2$ **f** $(50)^3$ **g** $(200)^2$ **h** 40×150
  i 60×5000 **j** 30×250 **k** 700×200

4 Evaluate the following.
  a $4000 \div 800$ **b** $9000 \div 30$ **c** $7000 \div 200$ **d** $8000 \div 200$
  e $2100 \div 700$ **f** $9000 \div 60$ **g** $700 \div 50$ **h** $3500 \div 70$
  i $3000 \div 500$ **j** $30\,000 \div 2000$ **k** $5600 \div 1400$ **l** $6000 \div 30$

5 Evaluate the following.
  a 7.3×10^2 **b** 3.29×10^5 **c** 7.94×10^3 **d** 6.8×10^7
  e $3.46 \div 10^2$ **f** $5.07 \div 10^4$ **g** $2.3 \div 10^4$ **h** $0.89 \div 10^3$

AU **6** The diameter of the planet Venus is approximately 7.5×10^3 miles.
  The diameter of Saturn is approximately 7.5×10^4 miles.
  The diameter of Earth is approximately 7.9×10^3 miles.
  Without working out the answers, explain how you can tell which planet is the biggest of the three.

PS **7** A number is between 10 000 and 100 000. It is written in the form 2.5×10^n.
  What is the value of n?

HOMEWORK 9G

1 Write these standard form numbers out in full.

   a 3.5×10^2 **b** 4.15×10 **c** 5.7×10^{-3} **d** 1.46×10

   e 3.89×10^{-2} **f** 4.6×10^3 **g** 2.7×10^2 **h** 8.6×10

   i 4.6×10^3 **j** 3.97×10^5 **k** 3.65×10^{-3} **l** 7.05×10^2

2 Write these numbers in standard form.

   a 780   **b** 0.435   **c** 67 800   **d** 7 400 000 000

   e 30 780 000 000   **f** 0.000 427 8   **g** 6450   **h** 0.047

   i 0.000 12   **j** 96.43   **k** 74.78   **l** 0.004 157 8

3 Write the appropriate numbers given in each statement in standard form.

   a In 1990 there were 24 673 000 vehicles licensed in the UK.

   b In 2001 Keith Gordon was one of 15 282 runners to complete the Boston Marathon.

   c In 1990 the total number of passenger kilometres on the British roads was 613 000 000 000.

   d The Sun is 93 million miles away from Earth. The next nearest star to the Earth is Proxima Centuri which is about 24 million million miles away.

   e A scientist was working with a new particle reported to weigh only 0.000 000 000 000 65 g.

AU 4 How many times smaller is 1.7×10^2 than 1.7×10^5?

AU 5 How many times bigger is 9.6×10^7 than 4.8×10^3?

PS 6 How many times bigger is 1.2×10^5 than 3000?

PS 7 The speed of light is 3.00×10^8 m/s ($2.997\ 924\ 58 \times 10^8$ m/s to be exact).
It takes about 1.3 seconds for light to travel to the moon.
Use this information to work out the distance to the moon.
Give your answer in kilometres.

HOMEWORK 9H

1 Find the results of the following, leaving your answers in standard form.

   a $(4 \times 10^6) \times (7 \times 10^9)$ **b** $(7 \times 10^5) \times (5 \times 10^7)$

   c $(3 \times 10^{-5}) \times (8 \times 10^8)$ **d** $(2.1 \times 10^7) \times (5 \times 10^{-8})$

2 Find the results of the following, leaving your answers in standard form.

   a $(9 \times 10^8) \div (3 \times 10^4)$ **b** $(2.7 \times 10^7) \div (9 \times 10^3)$

   c $(5.5 \times 10^4) \div (1.1 \times 10^{-2})$ **d** $(4.2 \times 10^{-9}) \div (3 \times 10^{-8})$

3 Find the results of the following, leaving your answers in standard form.

   a $\dfrac{8 \times 10^9}{4 \times 10^7}$ **b** $\dfrac{12 \times 10^6}{3 \times 10^4}$ **c** $\dfrac{2.8 \times 10^7}{7 \times 10^{-4}}$

4 $p = 8 \times 10^5$ and $q = 2 \times 10^7$

   Find the value of the following, leaving your answer in standard form.

   a $p \times q$ **b** $p \div q$ **c** $p + q$ **d** $q - p$ **e** $\dfrac{q}{p}$

5 $p = 2 \times 10^{-2}$ and $q = 4 \times 10^{-3}$

    a $p \times q$ **b** $p \div q$ **c** $p + q$ **d** $q - p$ **e** $\dfrac{q}{p}$

FM 6 In 2010 the population of Africa was approximately 1×10^9.

In 2050 the population of Africa is expected to be 1.8×10^9.

By how much is the population of Africa expected to rise?

Give your answer in millions.

AU 7 A number, when written in standard form, is greater than 1 million and less than 5 million.

Write down a possible value of the number in standard form.

PS 8 Here are four numbers written in standard form.

   3.5×10^5 1.2×10^3 7.3×10^2 4.8×10^4

    a Work out the largest answer when two of these numbers are multiplied together.

    b Work out the smallest answer when two of these numbers are added together.

Give your answers in standard form.

9.3 Rational numbers and reciprocals

HOMEWORK 9I

1 Work out each of these fractions as a decimal. Give them as terminating decimals or recurring decimals as appropriate.

    a $\frac{3}{4}$ **b** $\frac{1}{15}$ **c** $\frac{1}{25}$ **d** $\frac{1}{11}$ **e** $\frac{1}{20}$

PS 2 There are several patterns to be found in recurring decimals. For example,

    $\frac{1}{13} = 0.076923076923076923076923076230\ldots$; $\frac{2}{13} = 0.15384615384615384615384615846\ldots$

    $\frac{3}{13} = 0.230769230769230769230769\ldots$ and so on

    a Write down the decimals for $\frac{4}{13}, \frac{5}{13}, \frac{6}{13}, \frac{7}{13}, \frac{8}{13}, \frac{9}{13}, \frac{10}{13}, \frac{11}{13}, \frac{12}{13}$ to 24 decimal places.

    b What do you notice?

3 Write each of these fractions as a decimal. Use this to write the list in order of size, smallest first.

    $\frac{2}{9}$ $\frac{1}{5}$ $\frac{23}{100}$ $\frac{2}{7}$ $\frac{3}{11}$

4 Convert each of these terminating decimals to a fraction.

    a 0.57   **b** 0.275   **c** 0.85   **d** 0.06   **e** 3.65

AU 5 Explain why the reciprocal of 1 is 1.

PS 6   **a** Work out the reciprocal of the reciprocal of 4.

    b Work out the reciprocal of the reciprocal of 5.

    c What do you notice?

7   **a** Give an example to show that the reciprocal of a number greater than 1 is less than 1.

    b Give an example to show that the reciprocal of a number less than 0 is also less than 0.

8 $x = 0.0242\,424\ldots$

    a What is $100x$?

    b By subtracting the original value from your answer to part **a**, work out the value of $99x$.

    c Multiply both sides by 10 to get $990x$ and eliminate the decimal on the right-hand side.

    d Divide both sides by 990.

    e What is x as a fraction expressed in its lowest terms?

9 Convert each of these recurring decimals to a fraction.
  a $0.\dot{7}$ b $0.5\dot{7}$ c $0.5\dot{4}$ d $0.\dot{2}7\dot{5}$
  e $2.\dot{5}$ f $2.\dot{3}\dot{6}$ g $0.06\dot{3}$ h $2.07\dot{5}$

10 a Write 1.7 as a rational number in the form $\frac{a}{b}$, where a and b are whole numbers.
   b Given that $n = 1.\dot{7}$
     i   Write down the value of $10n$.
     ii   Hence write down the value of $9n$.
     iii Express n as a rational number, in the form $\frac{a}{b}$, where a and b are whole numbers.

9.4 Surds

HOMEWORK 9J

1 Work out each of the following in simplified form.
  a $\sqrt{3} \times \sqrt{4}$ b $\sqrt{5} \times \sqrt{7}$ c $\sqrt{5} \times \sqrt{5}$ d $\sqrt{2} \times \sqrt{32}$

PS 2 Work out each of the following in surd form.
  a $\sqrt{15} \div \sqrt{5}$ b $\sqrt{18} \div \sqrt{2}$ c $\sqrt{32} \div \sqrt{2}$ d $\sqrt{12} \div \sqrt{8}$

3 Work out each of the following in surd form.
  a $\sqrt{3} \times \sqrt{3} \times \sqrt{2}$ b $\sqrt{5} \times \sqrt{5} \times \sqrt{15}$ c $\sqrt{2} \times \sqrt{8} \times \sqrt{8}$ d $\sqrt{2} \times \sqrt{8} \times \sqrt{5}$

4 Work out each of the following in surd form.
  a $\sqrt{3} \times \sqrt{8} \div \sqrt{2}$ b $\sqrt{15} \times \sqrt{3} \div \sqrt{5}$ c $\sqrt{8} \times \sqrt{8} \div \sqrt{2}$ d $\sqrt{3} \times \sqrt{27} \div \sqrt{3}$

5 Simplify each of the following surds into the form $a\sqrt{b}$.
  a $\sqrt{90}$ b $\sqrt{32}$ c $\sqrt{63}$ d $\sqrt{300}$
  e $\sqrt{150}$ f $\sqrt{270}$ g $\sqrt{96}$ h $\sqrt{125}$

6 Simplify each of these.
  a $2\sqrt{32} \times 5\sqrt{2}$ b $4\sqrt{8} \times 2\sqrt{2}$ c $4\sqrt{12} \times 5\sqrt{3}$ d $3\sqrt{6} \times 2\sqrt{6}$
  e $2\sqrt{5} \times 5\sqrt{3}$ f $2\sqrt{3} \times 3\sqrt{3}$ g $2\sqrt{2} \times 3\sqrt{8}$ h $2\sqrt{3} \times 2\sqrt{27}$
  i $8\sqrt{24} \div 2\sqrt{3}$ j $3\sqrt{27} \div \sqrt{3}$ k $5\sqrt{18} \div \sqrt{2}$ l $2\sqrt{32} \div 4\sqrt{8}$
  m $5\sqrt{2} \times \sqrt{8} \div 2\sqrt{2}$ n $3\sqrt{15} \times \sqrt{3} \div \sqrt{5}$ o $2\sqrt{24} \times 5\sqrt{3} \div 2\sqrt{8}$

PS 7 Find the value of a that makes each of these surds true.
  a $\sqrt{5} \times \sqrt{a} = 20$ b $\sqrt{3} \times \sqrt{a} = 12$ c $\sqrt{5} \times 4\sqrt{a} = 20$

8 Simplify the following.
  a $\left(\frac{\sqrt{2}}{3}\right)^2$ b $\left(\frac{4}{\sqrt{3}}\right)^2$

9 Simplify the following.
  a $\sqrt{32} + \sqrt{8}$ b $\sqrt{32} \times \sqrt{8}$ c $\sqrt{27} \times \sqrt{18} \div \sqrt{3}$

AU 10 Decide whether this statement is true or false.
  $\sqrt{(a^2 + b^2)} = a + b$
  Show your working.

PS 11 Write down a division of two different surds which has an integer answer.

HOMEWORK 9K

A*

1 Rationalise the denominators of these expressions.

a $\dfrac{1}{\sqrt{7}}$ b $\dfrac{1}{\sqrt{8}}$ c $\dfrac{2}{\sqrt{5}}$ d $\dfrac{1}{2\sqrt{2}}$

e $\dfrac{5\sqrt{3}}{\sqrt{27}}$ f $\dfrac{\sqrt{8}}{\sqrt{3}}$ g $\dfrac{1+\sqrt{3}}{\sqrt{3}}$ h $\dfrac{3-\sqrt{2}}{\sqrt{8}}$

2 Show that:

a $(3+\sqrt{5})(2+\sqrt{5}) = 11 + 5\sqrt{5}$ b $(3-\sqrt{2})(3+\sqrt{2}) = 7$

3 Expand and simplify where possible.

a $\sqrt{5}(3-\sqrt{2})$ b $\sqrt{8}(3-4\sqrt{2})$ c $3\sqrt{8}(2\sqrt{2}+4)$
d $(2+\sqrt{3})(1-\sqrt{3})$ e $(3+\sqrt{5})(2-\sqrt{5})$ f $(3-\sqrt{2})(4+2\sqrt{2})$

4 Work out the missing lengths in these triangles, simplifying the answer where possible.

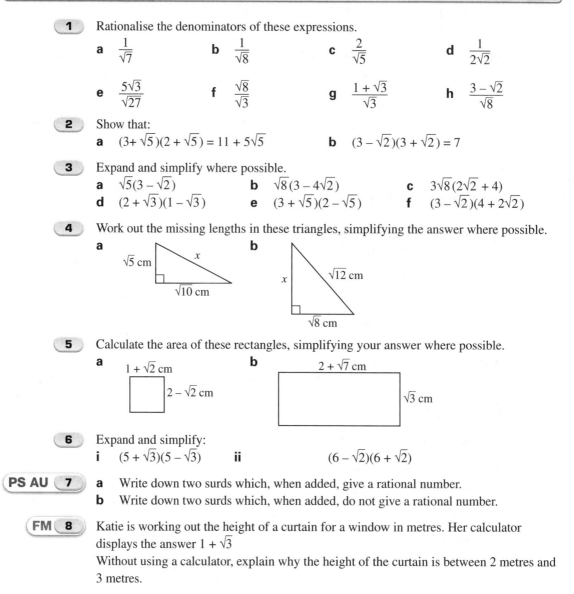

5 Calculate the area of these rectangles, simplifying your answer where possible.

6 Expand and simplify:

i $(5+\sqrt{3})(5-\sqrt{3})$ ii $(6-\sqrt{2})(6+\sqrt{2})$

PS AU 7 a Write down two surds which, when added, give a rational number.
b Write down two surds which, when added, do not give a rational number.

FM 8 Katie is working out the height of a curtain for a window in metres. Her calculator displays the answer $1 + \sqrt{3}$
Without using a calculator, explain why the height of the curtain is between 2 metres and 3 metres.

Functional Maths Activity

The planets

The table shows information about the planets in our solar system in order of distance from the sun.

Planet	Distance from the sun (million km)	Mass (kg)	Diameter (km)
Mercury	58	3.3×10^{23}	4878
Venus	108	4.87×10^{24}	12 104
Earth	150	5.98×10^{24}	12 756
Mars	228	6.42×10^{23}	6787
Jupiter	778	1.90×10^{27}	142 796
Saturn	1427	5.69×10^{26}	120 660
Uranus	2871	8.68×10^{25}	51 118
Neptune	4497	1.02×10^{26}	48 600
Pluto	5913	1.29×10^{22}	2274

Task 1

Answer the following questions.

1 Which planet is the largest?
2 Which planet is the smallest?
3 Which planet is the lightest?
4 Which planet is the heaviest?
5 Which planet is approximately twice as far away from the sun as Saturn?
6 Which two planets are similar in size?

Task 2

Sort the planets into order, lightest to heaviest, by mass.

Task 3

Sort the planets into order, shortest to longest, by diameter.

Task 4

Imagine you are a scientist working for NASA. Can you identify a relationship between each planet's mass, distance from the sun and diameter? Are there any trends? Write your findings in a brief report.

Algebra: Patterns

10.1 Number sequences

HOMEWORK 10A

1 Look at the following number sequences. Write down the next three terms in each and try to explain the pattern.

    a   4, 6, 8, 10, …   **b**   3, 6, 9, 12, …   **c**   2, 4, 8, 16, …

    d   5, 12, 19, 26, …   **e**   3, 30, 300, 3000, …   **f**   1, 4, 9, 16, …

2 Look carefully at each number sequence below. Find the next two numbers in the sequence and try to explain the pattern.

    a   1, 1, 2, 3, 5, 8, 13, 21, …   **b**   2, 3, 5, 8, 12, 17, …

3 Look at the sequences below. Find the rule for each sequence and write down its next three terms.

    a   7, 14, 28, 56, …   **b**   3, 10, 17, 24, 31, …   **c**   1, 3, 7, 15, 31, …

    d   40, 39, 37, 34, …   **e**   3, 6, 11, 18, 27, …   **f**   4, 5, 7, 10, 14, 19, …

    g   4, 6, 7, 9, 10, 12, …   **h**   5, 8, 11, 14, 17, …   **i**   5, 7, 10, 14, 19, 25, …

    j   10, 9, 7, 4, …   **k**   200, 40, 8, 1.6, …   **l**   3, 1.5, 0.75, 0.375, …

FM 4 A physiotherapist uses the formula below for charging for a series of n sessions when they are paid for in advance.

For $n \leqslant 5$, cost will be £$(35n + 20)$

For $6 \leqslant n \leqslant 10$, cost will be £$(35n + 10)$

For $n \geqslant 11$, cost will be £$35n$

    a   How much will the physiotherapist charge for eight sessions booked in advance?

    b   How much will the physiotherapist charge for 14 sessions booked in advance?

    c   One client paid £220 in advance for a series of sessions.

         How many sessions did she book?

    d   A runner has a leg injury and is not sure how many sessions it will take to cure. The runner books four sessions in advance, and after the sessions starts to run in races again. The leg injury returns and he has to book another three sessions before he is finally cured.

         How much more did it cost him than if he had booked seven sessions in advance?

PS 5 The formula for working out a series of fractions is

$$\frac{n + 2}{2n + 1}$$

Show that in the first eight terms only one of the fractions is a terminating decimal.

> **HINTS AND TIPS**
>
> If you set this up on a spreadsheet, find the relationship between the denominators of the terms that give terminating decimals in this series.

   FM Functional Maths   **AU** (AO2) Assessing Understanding   **PS** (AO3) Problem Solving

10.2 Finding the nth term of a linear sequence

HOMEWORK 10B

1 Find the nth term in each of these linear sequences.
   a 5, 7, 9, 11, 13 …
   b 3, 11, 19, 27, 35, …
   c 6, 11, 16, 21, 26, …
   d 3, 9, 15, 21, 27, …
   e 4, 7, 10, 13, 16, …
   f 3, 10, 17, 24, 31, …

2 Find the 50th term in each of these linear sequences.
   a 3, 5, 7, 9, 11, …
   b 5, 9, 13, 17, 21, …
   c 8, 13, 18, 23, 28, …
   d 2, 8, 14, 20, 26, …
   e 5, 8, 11, 14, 17, …
   f 2, 9, 16, 23, 30, …

3 For each sequence **a** to **f**, find
   i the nth term   **ii** the 100th term   **iii** the term closest to 100
   a 4, 7, 10, 13, 16, …
   b 7, 9, 11, 13, 15, …
   c 3, 8, 13, 18, 23, …
   d 1, 5, 9, 13, 17, …
   e 2, 10, 18, 26, …
   f 5, 6, 7, 8, 9, …

4 A sequence of fractions is $\frac{3}{5}, \frac{5}{8}, \frac{7}{11}, \frac{9}{14}, \ldots$
   a Find the nth term in the sequence.   **b** Change each fraction to a decimal.
   c What, as a decimal, will be the value of the:
      i 100th term   **ii** 1000th term?
   d Use your answers to part **c** to predict what the 10 000th term and the millionth term are. (Check these out on your calculator.)

5 **a** A number pattern begins 1, 1, 2, 3, 5, 8, …
      i What is the next number in this pattern?
      ii The number pattern is continued. Explain how you would find the 10th number in the pattern.
   b Another number pattern begins 1, 5, 9, 13, 17, … . Write down, in terms of n, the nth term in this pattern.

FM 6 This chart is used by a taxi firm for the charges for journeys of k kilometres.

k	1	2	3	4	5	6	7	8	9	10
Charge (£)	4.50	6.50	8.50	10.50	12.50	15.00	17.00	19.00	21.00	23.00
k	11	12	13	14	15	16	17	18	19	20
Charge (£)	26.00	28.00	30.00	32.00	34.00	37.00	39.00	41.00	43.00	45.00

   a Using the charges for 1 to 5 km, work out an expression for the kth term.
   b Using the charges for 6 to 10 km, work out an expression for the kth term.
   c Using the charges for 10 to 15 km, work out an expression for the kth term.
   d Using the charges for 16 to 20 km, work out an expression for the kth term.
   e What is the basic charge per kilometre?

PS 7 A series of fractions is $\frac{3}{7}, \frac{5}{10}, \frac{7}{13}, \frac{9}{16}, \frac{11}{19}, \ldots$
   a Write down an expression for the nth term of the numerators.
   b Write down an expression for the nth term of the denominators.
   c **i** Work out the fraction when $n = 1000$.
      ii Give the answer as a decimal.
   d Will the terms of the series ever be greater than $\frac{2}{3}$?
      Explain your answer.

10.3 Special sequences

HOMEWORK 10C

PS 1 p is an even number, q is a square number.
State if the following are odd or even or could be either odd or even.

a $p + 1$ **b** $p + q$ **c** $2q$ **d** $3p - 1$
e p^2 **f** $pq - 1$ **g** $p^2 + 4q$ **h** $(p + q)(p - q)$

2 The powers of 3 are $3^1, 3^2, 3^3, 3^4, \ldots\ldots$
This gives the sequence $3, 9, 27, 81, \ldots\ldots$

a Continue the sequence for another three terms.
b The nth term is given by 3^n.
Give the nth terms of each of these sequences.
  i $2, 8, 26, 80, \ldots\ldots$
  ii $6, 18, 54, 162, \ldots\ldots$

3 The negative powers of 10, starting with the power 0, are:
$10^0, 10^{-1}, 10^{-2}, 10^{-3}, \ldots\ldots$
This gives the sequence $0, 0.1, 0.01, 0.001$

a Describe the connection between the numerical value of the power and the number of zeros after the decimal point.
b If $10^{-n} = 0.0000001$, what is the value of n?

4 **a** Complete the table below to show the possible outcomes of adding a prime number to odd and even numbers, showing whether the answer is always odd, always even or could be either odd or even.

+	Prime	Odd	Even
Prime	Either		
Odd		Even	
Even			Even

b Complete the table below to show the possible outcomes of multiplying a prime number to odd and even numbers, showing whether the answer is always odd, always even or could be either odd or even.

+	Prime	Odd	Even
Prime	Either		
Odd		Odd	
Even			Even

PS 5 **a** Draw an equilateral triangle with each side 9 cm.
The perimeter will be 27 cm.
b Draw another equilateral triangle of side 3 cm on each edge.
Work out the perimeter of the new shape.
c Draw another equilateral triangle of side 1 cm on each of the remaining edges.
Work out the perimeter of the new shape.
d The next step would be to draw a triangle of side $\frac{1}{3}$ cm on each remaining edge, but this will be difficult to draw.
You should be able to write down the perimeter using the pattern of the perimeters so far.

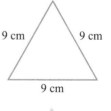

9 cm 9 cm

9 cm

3 cm

3 cm 3 cm

The formula is $27 \times \left(\dfrac{4}{3}\right)^{n-1}$

You will need a calculator with a power button (^).

Work out $27 \times (4 \div 3)$ ^ 0, which should equal 27.

Then work out $27 \times (4 \div 3)$ ^ 1, which should equal your answer to the perimeter in part **b**.

Use the formula to work out the perimeter of the next drawing when $n = 4$

e Work out the perimeter when $n = 100$

If we kept on drawing triangles, the perimeter would become infinite.

This is an example of a shape that has a finite area surrounded by an infinite perimeter.

10.4 General rules from given patterns

HOMEWORK 10D

 A pattern of shapes is built up from matchsticks as shown.

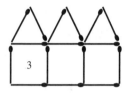

a Draw the fourth diagram.

b How many matchsticks are in the *n*th diagram?

c How many matchsticks are in the 25th diagram?

d With 200 matchsticks, which is the biggest diagram that could be made?

2 A pattern of hexagons is built up from matchsticks.

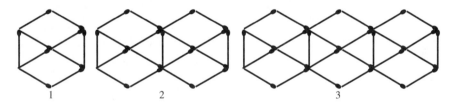

a Draw the fourth set of hexagons in this pattern.

b How many matchsticks are needed for the *n*th set of hexagons?

c How many matchsticks are needed to make the 60th set of hexagons?

d If there are only 100 matchsticks, which is the largest set of hexagons that could be made?

 A conference centre had tables each of which could sit three people. When put together, the tables could seat people as shown.

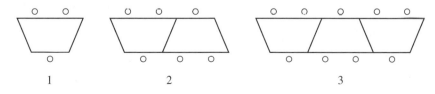

a How many people could be seated at four tables?

b How many people could be seated at *n* tables put together in this way?

c A conference had 50 people who wished to use the tables in this way. How many tables would they need?

PS **4** The picture shows a pattern of cards.

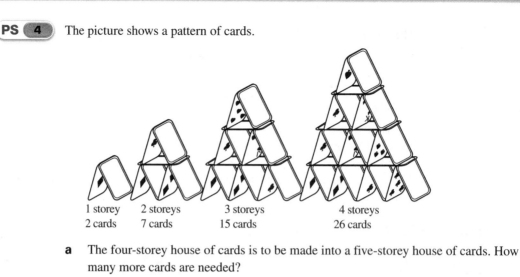

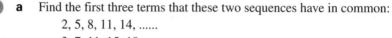

1 storey   2 storeys   3 storeys   4 storeys
2 cards   7 cards   15 cards   26 cards

a The four-storey house of cards is to be made into a five-storey house of cards. How many more cards are needed?

b Look at the sequence 2, 7, 15, 26, …
  i Calculate the sixth term in this sequence.
  ii Explain how you found your answer.

5 **a** Find the first three terms that these two sequences have in common:
  2, 5, 8, 11, 14, ...
  3, 7, 11, 15, 19, ...

b Write down the nth term of the sequence that is the answer to part **a**.

Problem-solving Activity

Packaging

To tie up a cuboidal package that is L cm long, W cm wide and H cm high, this formula gives the length of string needed:

$S = 2L + 2W + 4H + 20$

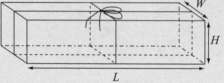

Masood wants to send eight identical cuboidal boxes with sides of 15 cm in one package.

He can arrange them in three different ways to make a cuboidal shape.

Two of these ways are shown below.

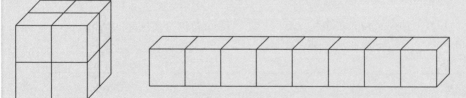

Which of the three ways of arranging the eight cubes in a cuboid will use the least amount of string?

Algebra: Expressions and equations

11.1 Basic algebra

D

1 Find the value of $4x + 3$ when **a** $x = 3$ **b** $x = 6$ **c** $x = 11$

2 Find the value of $3k - 1$ when **a** $k = 2$ **b** $k = 5$ **c** $k = 10$

3 Find the value of $4 + t$ when **a** $t = 5$ **b** $t = 8$ **c** $t = 15$

4 Evaluate $14 - 3f$ when **a** $f = 4$ **b** $f = 6$ **c** $f = 10$

5 Evaluate $\dfrac{4d - 7}{2}$ when **a** $d = 2$ **b** $d = 5$ **c** $d = 15$

6 Find the value of $5x + 2$ when **a** $x = -2$ **b** $x = -1$ **c** $x = 21.5$

7 Evaluate $4w - 3$ when **a** $w = -2$ **b** $w = -3$ **c** $w = 2.5$

8 Evaluate $10 - x$ when **a** $x = -3$ **b** $x = -6$ **c** $x = 4.6$

9 Find the value of $5t - 1$ when **a** $t = 2.4$ **b** $t = -2.6$ **c** $t = 0.05$

10 Evaluate $11 - 3t$ when **a** $t = 2.5$ **b** $t = -2.8$ **c** $t = 0.99$

11 Where $H = a^2 + c^2$, find H when **a** $a = 3$ and $c = 4$ **b** $a = 5$ and $c = 12$

12 Where $K = m^2 - n^2$, find K when **a** $m = 5$ and $n = 3$ **b** $m = -5$ and $n = -2$

13 Where $P = 100 - n^2$, find P when **a** $n = 7$ **b** $n = 8$ **c** $n = 9$

14 Where $D = 5x - y$, find D when **a** $x = 4$ and $y = 3$ **b** $x = 5$ and $y = -3$

15 Where $T = y(2x + 3y)$, find T when **a** $x = 8$ and $y = 12$ **b** $x = 5$ and $y = 7$

16 Where $m = w(t^2 + w^2)$, find m when **a** $t = 5$ and $w = 3$ **b** $t = 8$ and $w = 7$

FM 17 Two of the first recorded units of measurement were the *cubit* and the *palm*.
The cubit is the distance from the fingertip to the elbow and the palm is the distance across the hand.
A cubit is four and a half palms.
The actual length of a cubit varied throughout history, but it is now accepted to be 54 cm.
Noah's Ark is recorded as being 300 cubits long by 50 cubits wide by 30 cubits high.
What are the dimensions of the Ark in metres?

AU 18 In this algebraic magic square, every row, column and diagonal should add up and simplify to $9a + 6b + 3c$.

$3a - 3b + 4c$	$2a + 8b + c$	$4a + b - 2c$
	$3a + 2b + c$	$2a - 2b + 7c$
$2a + 3b + 4c$		$3a + 7b - 2c$

a Copy and complete the magic square.
b Calculate the value of the 'magic number' if $a = 2$, $b = 3$ and $c = 4$.

AU 19 The rule for converting degrees Fahrenheit into degrees Celsius is:
$$C = \frac{5}{9}(F - 32)$$

a Use this rule to convert 68 °F into degrees Celsius.
b Which of the following is the rule for converting degrees Celsius into degrees Fahrenheit?
$$F = \frac{9}{5}(C + 32) \qquad F = \frac{5}{9}C + 32 \qquad F = \frac{9}{5}C + 32 \qquad F = \frac{9}{5}C - 32$$

FM 20 The formula for the cost of water used by a household each quarter is:
£32.40 + £0.003 per litre of water used.
A family uses 450 litres of water each day.
a How much is their total bill per quarter? (Take a quarter to be 91 days.)
b The family pay a direct debit of £45 per month towards their electricity costs. By how much will they be in credit or debit after the quarter?

21 Using $x = 17.4$, $y = 28.2$ and $z = 0.6$, work out the value of:
a $x = \frac{y}{z}$ **b** $\frac{x + y}{z}$ **c** $\frac{x}{z} + y$

22 a Laser printer cartridges cost £75 and print approximately 2500 pages. Approximately how many pence per page does it cost to run, taking only ink consumption into consideration?
b A printing specialist uses a laser printer of this type. He charges a fixed rate of £4.50 to set up the design and five pence for every page. Explain why his profit on a print run of x pages is, in pounds, $4.5 + 0.02x$
c How much profit will the printing specialist make if he prints 2000 race entry forms for a running club?

HOMEWORK 11B

Expand these expressions.

1	$3(4 + m)$	**2**	$6(3 + p)$	**3**	$4(4 - y)$	**4**	$3(6 + 7k)$
5	$4(3 - 5f)$	**6**	$2(4 - 23w)$	**7**	$7(g + h)$	**8**	$4(2k + 4m)$
9	$6(2d - n)$	**10**	$t(t + 5)$	**11**	$m(m + 4)$	**12**	$k(k - 2)$
13	$g(4g + 1)$	**14**	$y(3y - 21)$	**15**	$p(7 - 8p)$	**16**	$2m(m + 5)$
17	$3t(t - 2)$	**18**	$3k(5 - k)$	**19**	$2g(4g + 3)$	**20**	$4h(2h - 3)$
21	$2t(6 - 5t)$	**22**	$4d(3d + 5e)$	**23**	$3y(4y + 5k)$	**24**	$6m^2(3m - p)$
25	$y(y^2 + 7)$	**26**	$h(h^3 + 9)$	**27**	$k(k^2 - 4)$	**28**	$3t(t^2 + 3)$
29	$5h(h^3 - 2)$	**30**	$4g(g^3 - 3)$	**31**	$5m(2m^2 + m)$	**32**	$2d(4d^2 - d^3)$
33	$4w(3w^2 + t)$	**34**	$3a(5a^2 - b)$	**35**	$2p(7p^3 - 8m)$	**36**	$m^2(3 + 5m)$
37	$t^3(t + 3t)$	**38**	$g^2(4t - 3g^2)$	**39**	$2t^2(7t + m)$	**40**	$3h^2(4h + 5g)$

FM **41** An approximate rule for converting degrees Fahrenheit into degrees Celsius is:
$C = 0.5(F - 30)$
a Use this rule to convert 22 °F into degrees Celsius.
b Which of the following is an approximate rule for converting degrees Celsius into degrees Fahrenheit?
$F = 2(C + 30)$ $F = 0.5(C + 30)$ $F = 2(C + 15)$ $F = 2(C - 15)$

AU **42** Copy the diagram below and draw lines to show which algebraic expressions are equivalent. One line has been drawn for you.

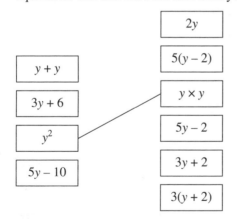

PS **43** The expansion $3(4x + 8y) = 12x + 24y$.
Write down two other expansions that give an answer of $12x + 24y$.

HOMEWORK 11C

1 Simplify these expressions.

a	$5t + 4t$	**b**	$4m + 3m$	**c**	$6y + y$	**d**	$2d + 3d + 5d$
e	$7e - 5e$	**f**	$6g - 3g$	**g**	$3p - p$	**h**	$5t - t$
i	$t^2 + 4t^2$	**j**	$5y^2 - 2y^2$	**k**	$4ab + 3ab$	**l**	$5a^2d - 4a^2d$

2 Expand and simplify.

a $3(2 + t) + 4(3 + t)$ **b** $6(2 + 3k) + 2(5 + 3k)$ **c** $5(2 + 4m) + 3(1 + 4m)$

d $3(4 + y) + 5(1 + 2y)$ **e** $5(2 + 3f) + 3(6 - f)$ **f** $7(2 + 5g) + 2(3 - g)$

g $4(3 + 2h) - 2(5 + 3h)$ **h** $5(3g + 4) - 3(2g + 5)$ **i** $3(4y + 5) - 2(3y + 2)$

j $3(5t + 2) - 2(4t + 5)$ **k** $5(5k + 2) - 2(4k - 3)$ **l** $4(4e + 3) - 2(5e - 4)$

m $m(5 + p) + p(2 + m)$ **n** $k(4 + h) + h(5 + 2k)$ **o** $t(1 + 2n) + n(3 + 5t)$

p $p(5q + 1) + q(3p + 5)$ **q** $2h(3 + 4j) + 3j(h + 4)$ **r** $3y(4t + 5) + 2t(1 + 4y)$

s $t(2t + 5) + 2t(4 + t)$ **t** $3y(4 + 3y) + y(6y - 5)$ **u** $5w(3w + 2) + 4w(3 - w)$

v $4p(2p + 3) - 3p(2 - 3p)$ **w** $4m(m - 1) + 3m(4 - m)$ **x** $5d(3 - d) + d(2d - 1)$

y $5a(3b + 2a) + a(2a^2 + 3c)$ **z** $4y(3w + y^2) + y(3y - 4t)$

FM 3 Adult tickets for a concert cost £x and children's tickets cost £y.

At the afternoon show there were 40 adults and 160 children.

At the evening show there were 60 adults and 140 children.

a Write down an expression for the total amount of money taken on that day in terms of x and y.

b The daily expense for putting on the show is £2200. If $x = 12$ and $y = 9$, how much profit did the theatre make that day?

AU 4 Don wrote the following:

$2(3x - 1) + 5(2x + 3) = 5x - 2 + 10x + 15 = 15x - 13$

Don has made two mistakes in his working.

Explain the mistakes that Don has made.

PS 5 An internet site sells CDs. They cost £$(x + 0.75)$ each for the first five and then £$(x + 0.25)$ for any orders over five.

a Moe buys eight CDs. Which of the following expressions represents how much Moe will pay?

i $8(x + 0.75)$ **ii** $5(x + 0.75) + 3(x + 0.25)$

iii $3(x + 0.75) + 5(x + 0.25)$ **iv** $8(x + 0.25)$

b If $x = 5$, how much will Moe pay?

11.2 Factorisation

HOMEWORK 11D

Factorise the following expressions.

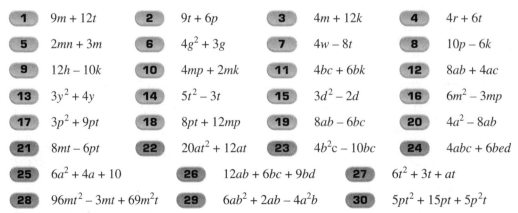

1 $9m + 12t$ **2** $9t + 6p$ **3** $4m + 12k$ **4** $4r + 6t$

5 $2mn + 3m$ **6** $4g^2 + 3g$ **7** $4w - 8t$ **8** $10p - 6k$

9 $12h - 10k$ **10** $4mp + 2mk$ **11** $4bc + 6bk$ **12** $8ab + 4ac$

13 $3y^2 + 4y$ **14** $5t^2 - 3t$ **15** $3d^2 - 2d$ **16** $6m^2 - 3mp$

17 $3p^2 + 9pt$ **18** $8pt + 12mp$ **19** $8ab - 6bc$ **20** $4a^2 - 8ab$

21 $8mt - 6pt$ **22** $20at^2 + 12at$ **23** $4b^2c - 10bc$ **24** $4abc + 6bed$

25 $6a^2 + 4a + 10$ **26** $12ab + 6bc + 9bd$ **27** $6t^2 + 3t + at$

28 $96mt^2 - 3mt + 69m^2t$ **29** $6ab^2 + 2ab - 4a^2b$ **30** $5pt^2 + 15pt + 5p^2t$

31 Factorise the following expressions where possible. List those which cannot factorise.

a $5m - 6t$	**b** $3m + 2mp$	**c** $t^2 - 5t$	**d** $6pt + 5ab$
e $8m^2 - 6mp$	**f** $a^2 + c$	**g** $3a^2 - 7ab$	**h** $4ab + 5cd$
i $7ab - 4b^2c$	**j** $3p^2 - 4t^2$	**k** $6m^2t + 9t^2m$	**l** $5mt + 3pn$

FM 32 An ink cartridge is priced at £9.99.

The shop has a special offer of 20% off if you buy five or more.

20% of £9.99 is £1.99.

Tom wants six cartridges. Tess wants eight cartridges.

Tom writes down the calculation $6 \times 9.99 - 6 \times 1.99$ to work out how much he must pay.

Tess writes down the calculation $8 \times (9.99 - 1.99)$ to work out how much she must pay.

Both calculations are correct.

a Who has the easier calculation and why?

b How much will each of them pay for their cartridges?

AU 33 **a** Factorise these expressions.

  i $4x + 3 + 5x - 7 - 8x$

  ii $3x - 12$

  iii $x^2 - 4x$

b What do all the answers in **a** have in common?

PS 34 To keep a class quiet a cover teacher asked them to add up all the numbers from 1 to 100 (i.e. $1 + 2 + 3 + 4 + \ldots + 98 + 99 + 100$).

Two minutes later, a student said she had the correct answer.

The teacher asked the student to show the class her method.

The student wrote:

$(1 + 100) + (2 + 99) + (3 + 98) + \ldots (50 + 51) = 50 \times 101$

a Explain why this gives the correct answer.

b What is the sum of all the numbers from 1 to 100?

11.3 Rearranging formulae

HOMEWORK 11E

FM 1 A restaurant has a large oven that can cook up to 10 chickens at a time.

The restaurant uses the following formula for the length of time it takes to cook n chickens:

$T = 10n + 55$

A large party is booked for a chicken dinner at 7 pm. They will need eight chickens between them.

a It takes 15 minutes to get the chickens out of the oven and prepare them for serving. At what time should the eight chickens go into the oven?

b The next day another large party is booked.

  i Rearrange the formula to make n the subject.

  ii The party is booked for 8 pm and the chef calculates she will need to put the chickens in the oven at 5.50 pm.

  How many chickens do the party need?

AU 2 Kern notices that the price of six coffees is 90 pence less than the price of nine teas.

Let the price of a coffee be x pence and the price of a tea be y pence.

a Express the cost of a tea, y, in terms of the price of a coffee, x.

b If the price of a coffee is £1.20, how much is a tea?

PS 3 Distance, speed and time are connected by the formula:

Distance = Speed $\leftrightarrow$ Time

A delivery driver drove 90 miles at an average speed of 60 miles per hour.
On the return journey, he was held up at some road works for 30 minutes.
What was his average speed on the return journey?

4 $y = mx + c$ **a** Make c the subject.   **b** Express x in terms of y, m and c.

5 $v = u - 10t$ **a** Make u the subject.   **b** Express t in terms of v and u.

6 $T = 2x + 3y$ **a** Express x in terms of T and y.   **b** Make y the subject.

7 $p = q^2$ Make q the subject.

8 $p = q^2 - 3$ Make q the subject.

9 $a = b^2 + c$ Make b the subject.

10 A rocket is fired vertically upwards with an initial velocity of u metres per second. After t seconds the rocket's velocity, v metres per second, is given by the formula $v = u + gt$, where g is a constant.

   a   Calculate v when $u = 120$, $g = -9.8$ and $t = 6$

   b   Rearrange the formula to express t in terms of v, u, and g.

   c   Calculate t when $u = 100$, $g = -9.8$ and $v = 17.8$

Problem-solving Activity

Eating out

Six people pay the following for their meals:

Ann:	Burger, Chips, Beans	£3.30
Bashir:	Sausage, Chips and Beans	£3.00
Carol:	Burger, Roast Potatoes and Peas	£3.30
Derek:	Burger, Chips and Peas	£3.40
Emir:	Sausage and Roast Potatoes	£2.30
Farook:	Burger and Peas	£2.50

Use this information to work out how much each item costs.

Algebra: Real-life graphs

12.1 Straight-line distance–time graphs

HOMEWORK 12A

FM 1 Joe was travelling in his car to meet his girlfriend. He set off from home at 9.00 am, and stopped on the way for a break. This distance–time graph illustrates his journey.

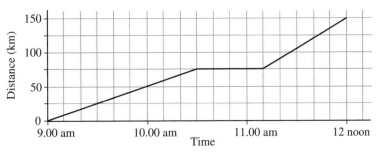

a At what time did he:
  i stop for his break   **ii** set off after his break   **iii** get to his meeting place?
b At what average speed was he travelling:
  i over the first hour   **ii** over the last hour   **iii** for his whole journey?

FM 2 A taxi set off from Hellaby to pick up Jean. It then went on to pick up Jean's parents. It then travelled further, dropping them all off at a shopping centre. The taxi travelled a further 10 km to pick up another party and took them back to Hellaby. This distance–time graph illustrates the journey.

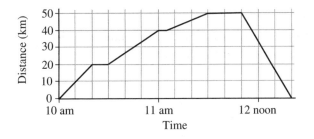

a How far from Hellaby did Jean's parents live?
b How far from Hellaby is the shopping centre?
c What was the average speed of the taxi while only Jean was in the taxi?
d What was the average speed of the taxi back to Hellaby?

FM 3 Grandad took his grandchildren out for a trip. He set off at 1 pm and travelled, for half an hour, away from Sheffield at the average speed of 60 km/h. They stopped to look at the sea and have an ice cream. At two o'clock, they set off again, travelling for a quarter of an hour at a speed of 80 km/h. Here they stopped to play on the sand for half an hour. Grandad then drove the grandchildren back home, at an average speed of 50 km/h. Draw a travel graph to illustrate this story. Use a horizontal axis to represent the time from 1 pm to 5 pm, and a vertical scale from 0 km to 50 km.

AU 4 A runner sets off at 8 am from point P to jog along a trail at a steady pace of 12 km/h.

One hour later, a cyclist sets off from P on the same trail at a steady pace of 24 km/h. After 30 minutes, the cyclist gets a puncture, which takes 30 minutes to fix. She then sets of at a steady pace of 24 km/h. At what time did the cyclist catch up with the runner? You may use a grid to help you solve this question.

> **HINTS AND TIPS**
>
> This question can be done by many methods, but doing a distance-time graph is the easiest. Mark a grid with a horizontal axis from 8 am to 12 noon and the vertical axis as distance from 0 to 40. Draw lines for both runner and cyclist. Remember that the cyclist doesn't start until 9 am.

HOMEWORK 12B

1 Calculate the gradient of each line.

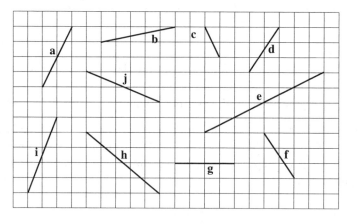

2 Calculate the average speed of the journey represented by each line in the following graphs. The gradient of each line is the speed.

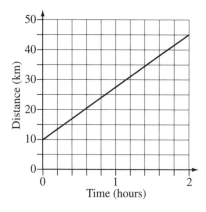

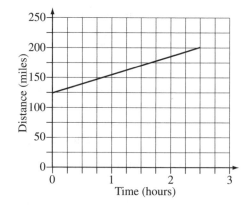

3 This is a conversion graph between ounces and grams.
  a Calculate the gradient of the line.
  b Use the graph to find the number of grams equivalent to 1 ounce.

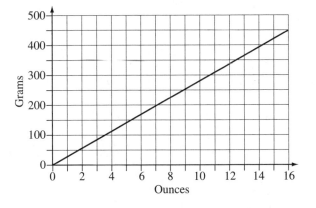

FM 4 This graph shows the journey of a car from London to Brighton and back again.

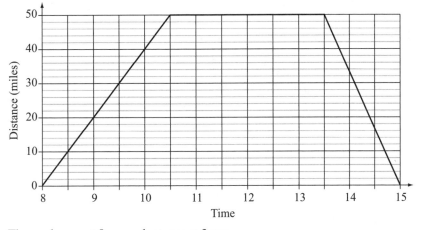

The car leaves at 8 am and returns at 3 pm.
  a For how long does the car stop in Brighton?
  b Was the car travelling faster on the way to Brighton or on the way back to London?
     Explain how you can tell this from the graph.

PS 5 The diagram shows the safe position for a ladder.
Use a grid like the one below and a protractor to work out the
approximate safe angle between the ladder and the ground,
marked x on the diagram.

12.2 Other types of graphs

HOMEWORK 12C

A

FM 1 The diagram shows the velocity of
a car over 10 seconds.
Calculate the acceleration
   a   over the first 2 seconds
   b   after 6 seconds.

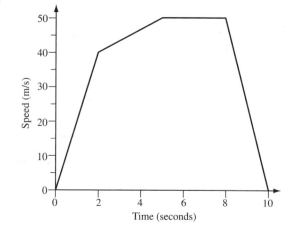

FM 2 The diagram shows the velocity–time graph for a short train journey between stops.
Find:
   a   the acceleration over the first 10 seconds
   b   the deceleration over the last 20 seconds.

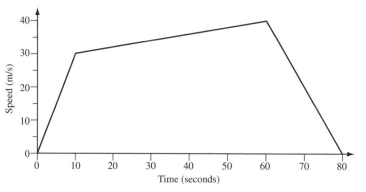

PS 3 Starting from rest (zero velocity), a car travels as indicated below.
   ● Accelerates at a constant rate over 10 seconds to reach 20 m/s.
   ● Keeps this velocity for 30 seconds.
   ● Accelerates over the next 10 seconds to reach 30 m/s.
   ● Steadily slows down to reach rest (zero velocity) over the next 20 seconds.
   a   Draw the velocity–time graph.
   b   Calculate the deceleration over the last 20 seconds.

PS **4** This graph represents the journey of a train.

a What was the average speed of the train from A to B?

b How long did the train wait at B?

c Another train starts from C at 11 am and travels non-stop to A at an average speed of 60 mph. Draw the graph of its journey on the copy of the graph.

d Write down how far from A the trains were when they passed each other.

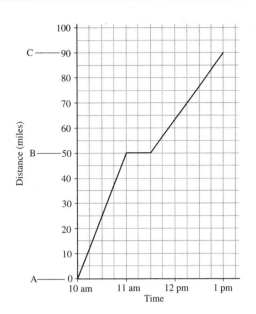

12.3 Linear graphs

HOMEWORK 12D

Draw the graph for each of the equations given.

Follow these hints.

● Use the highest and smallest values of x given as your range.

● When the first part of the function is a division, pick x-values that divide exactly to avoid fractions.

● Always label your graphs. This is particularly important when you are drawing two graphs on the same set of axes.

● Create a table of values. You will often have to complete these in your examinations.

1 Draw the graph of $y = 2x + 3$ for x-values from 0 to 5 ($0 \leqslant x \leqslant 5$)

2 Draw the graph of $y = 3x - 1$ ($0 \leqslant x \leqslant 5$)

3 Draw the graph of $y = \frac{x}{2} - 2$ ($0 \leqslant x \leqslant 12$)

4 Draw the graph of $y = 2x + 1$ ($-2 \leqslant x \leqslant 2$)

5 Draw the graph of $y = \frac{x}{2} + 5$ ($-6 \leqslant x \leqslant 6$)

6 a On the same set of axes, draw the graphs of $y = 3x - 1$ and $y = 2x + 3$ ($0 \leqslant x \leqslant 5$)

b Where do the two graphs cross?

7 a On the same axes, draw the graphs of $y = 4x - 3$ and $y = 3x + 2$ ($0 \leqslant x \leqslant 6$)

b Where do the two graphs cross?

D

8 **a** On the same axes, draw the graphs of

$$y = \frac{x}{2} + 1 \text{ and } y = \frac{x}{3} + 2 \quad (0 \leqslant x \leqslant 12)$$

b Where do the two graphs cross?

9 **a** On the same axes, draw the graphs of
$y = 2x + 3$ and $y = 2x - 1$ $(0 \leqslant x \leqslant 4)$

b Do the graphs cross? If not, why not?

10 **a** Copy and complete the table to draw the graph of $x + y = 6$ $(0 \leqslant x \leqslant 6)$

x	0	1	2	3	4	5	6
y							

b Now draw the graph of $x + y = 3$.

FM 11 CityCabs uses this formula to work out the cost of a journey of k kilometres:
$C = 2.5 + k$
TownCars uses this formula to work out the cost of a journey of k kilometres:
$C = 2 + 1.25k$

a On the grid below, draw lines to represent these formulae.

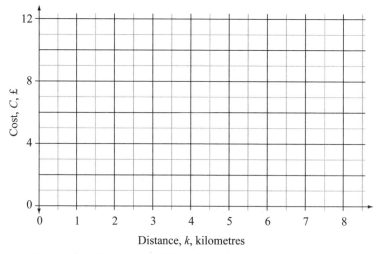

Distance, k, kilometres

b At what length of journey do CityCabs and TownCars charge the same amount?

AU 12 The line $x + y = 5$ is drawn on the grid below.

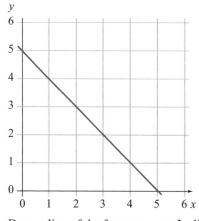

Draw a line of the form $x = a$ **and** a line of the form $y = b$ so that the area between the 3 lines is 4.5 square units.

HOMEWORK 12E

1 Find the gradient of each of these lines.

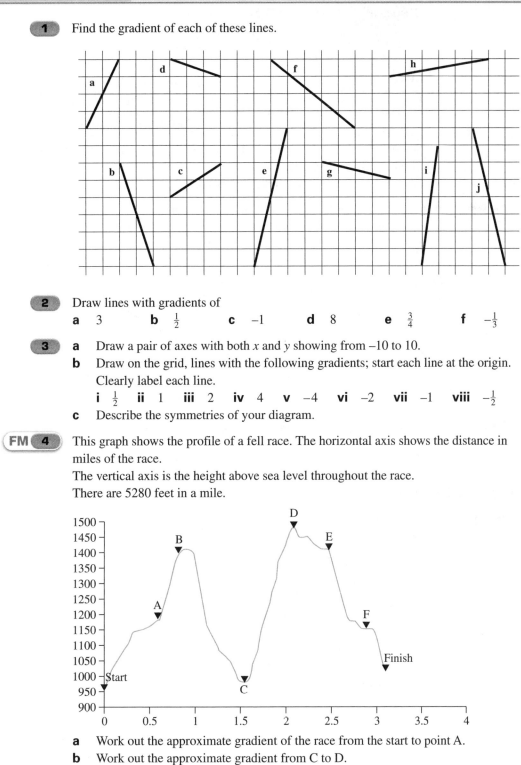

2 Draw lines with gradients of

a 3   **b** $\frac{1}{2}$ **c** −1   **d** 8   **e** $\frac{3}{4}$ **f** $-\frac{1}{3}$

3 **a** Draw a pair of axes with both x and y showing from −10 to 10.

b Draw on the grid, lines with the following gradients; start each line at the origin.
Clearly label each line.

i $\frac{1}{2}$ **ii** 1 **iii** 2 **iv** 4 **v** −4 **vi** −2 **vii** −1 **viii** $-\frac{1}{2}$

c Describe the symmetries of your diagram.

FM **4** This graph shows the profile of a fell race. The horizontal axis shows the distance in miles of the race.

The vertical axis is the height above sea level throughout the race.

There are 5280 feet in a mile.

a Work out the approximate gradient of the race from the start to point A.

b Work out the approximate gradient from C to D.

c Fell races are classified in terms of distance and amount of ascent.

Distance	S (Short)	Less than 6 miles
	M (Medium)	Between 6 and 12 miles
	L (Long)	Over 12 miles
Ascent	C	An average of 100 to 125 feet per mile
	B	An average of 125 to 250 feet per mile
	A	An average of 250 or more feet per mile

So, for example, an AL race would be over 12 miles and have at least 250 feet of ascent on average per mile.

What category is the race on the previous page?

AU 5 Write the gradients of the two lines below in the form $1 : n$.

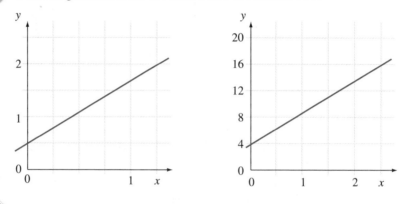

PS 6 Put the following gradients in order of steepness starting with the shallowest.

1 horizontal, 3 vertical   2 horizontal, 9 vertical   3 horizontal, 7 vertical

4 horizontal, 5 vertical   5 horizontal, 7 vertical   6 horizontal, 13 vertical

12.4 Uses of graphs

HOMEWORK 12F

1 This graph illustrates the charges made by an electricity company.

a Calculate the standing charge. This is the amount paid before any electricity is used.

b What is the gradient of the line?

c From your answers to **a** and **b** write down the rule to calculate the total charge for electricity.

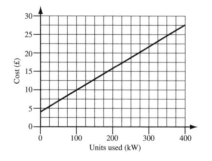

2 This graph illustrates the charges made by a gas company.
   a Calculate the standing charge.
   b What is the gradient of the line?
   c From your answers to **a** and **b** write down the rule to calculate the total charge for gas.

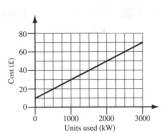

3 This graph illustrates the charges made by a phone company.
   a Calculate the standing charge.
   b What is the gradient of the line?
   c From your answers to **a** and **b** write down the rule to calculate the total charge for using this phone company.

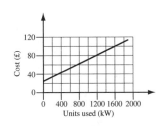

PS 4 The graph shows 3 line segments.

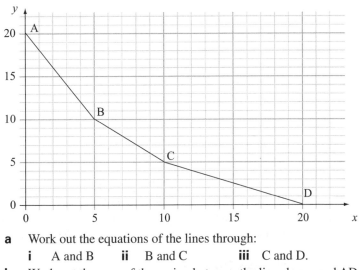

   a Work out the equations of the lines through:
     i A and B   **ii** B and C   **iii** C and D.
   b Work out the area of the region between the line shown and AD.

HOMEWORK 12G

By drawing their graphs, find the solution of each of these pairs of simultaneous equations.

1 $x + 4y = 1$
$x - y = 6$

2 $y = 2x + 1$
$3x + 2y = 23$

3 $y = 2x + 5$
$y = x + 4$

4 $y = x$
$x - y = 4$

5 $y + 10 = 2x$
$5x + y = 18$

6 $y = 5x - 1$
$y = 3x + 2$

7 $y = x + 11$
$x + y = 5$

8 $y - 3x = 8$
$y = x + 6$

9 $y = -x$
$y = 4x + 15$

10 $3x + 2y = 2$
$y = -2x$

11 $y = 3x - 4$
$y + x = 6$

12 $y = 3x - 12$
$x + y = 2$

B

AU **13** Two coffees and three cakes cost £7.00.
Two coffees and one cake cost £4.00.
Using x to represent the price of a coffee and y to represent the cost of a cake, set up a pair of simultaneous equations.
Using a set of axes with the x-axis from 0 to 4 and the y-axis from 0 to 4, draw the graphs of the two equations.
Use the graphs to write down the cost of a coffee and the cost of a cake.

PS **14** The graph shows four lines:
P: $y = -x$ Q: $y = 2x + 6$ R: $y = x - 2$ S: $y = -\frac{2}{3}x - 2$

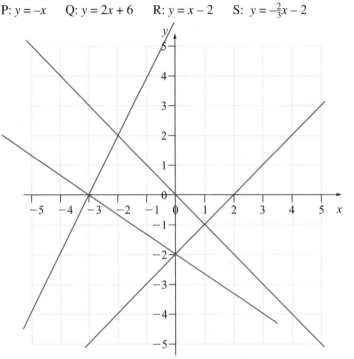

a   Which pair of lines intersect at the following points?
   i $(0, -2)$ **ii** $(-3, 0)$ **iii** $(1, -1)$ **iv** $(-2, 2)$
b   Solve the simultaneous equations given by P and S to find the exact solution.

12.5 Parallel and perpendicular lines

HOMEWORK 12H

A

1 Write down the equation of the line parallel to each of the following lines and which passes through point $(0, 1)$.
   a $y = 2x - 3$ **b** $y = -4x + 3$ **c** $y = \frac{1}{2}x - 5$ **d** $y = -\frac{1}{4}x - 3$

2 Write down the equations of these lines.
   a   parallel to $y = 3x - 2$ and passes through $(0, 4)$
   b   parallel to $y = \frac{1}{4}x + 3$ and passes through $(0, -1)$
   c   parallel to $y = -x + 3$ and passes through $(0, 2)$

3 Find the equation of the line that passes through $(4, 7)$ and is parallel to AB, where A is $(1, 4)$ and B is $(5, 2)$.

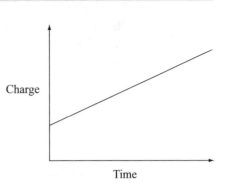

FM 4 Two taxi companies, Acabs and BeeCabs, use the same graph to charge for journeys. A sketch of this graph is shown here.

a Acabs decides to reduce its basic charge by 50% but to maintain the same charge per kilometre.
Sketch the new graph on a copy of the original graph.

b BeeCabs decide to have no basic charge but to double the charge per kilometre.
Sketch the new graph for BeeCabs on a copy of the original graph.

Charge

Time

12.6 Other graphs

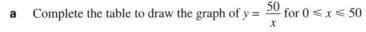
HOMEWORK 12I

1 a Complete the table to draw the graph of $y = x^3 + 1$ for $-3 \leqslant x \leqslant 3$

x	-3	-2	-1	0	1	2	3
$y = x^3 + 1$	-26			1			28

b Use your graph to find the y-value for an x-value of 1.2.

2 a Complete the table to draw the graph of $y = x^3 + 2x$ for $-2 \leqslant x \leqslant 3$

x	-2	-1	0	1	2	3
$y = x^3 + 2x$	-12		0		12	

b Use your graph to find the y-value for an x-value of 2.5.

3 a Complete the table to draw the graph of $y = \dfrac{12}{x}$ for $-12 \leqslant x \leqslant 12$

x	-12	-6	-4	-3	-2	-1	1	2	3	4	6	12
$y = \dfrac{12}{x}$	-1			-4					4			1

b Use your graph to find:
i the y-value when $x = 1.5$ **ii** the x-value when $y = 5.5$.

4 a Complete the table to draw the graph of $y = \dfrac{50}{x}$ for $0 \leqslant x \leqslant 50$

x	1	2	5	10	25	50
$y = \dfrac{50}{x}$						

b On the same axes, draw the line $y = x + 30$
c Use your graph to find the x-value of the point where the graphs cross.

5 a Complete the table below for $y = 2^x$ for values of x from -3 to $+4$. (Values are rounded to 2 dp.)

x	-3	-2	-1	0	1	2	3	4
$y = 2^x$	0.1	0.3			2	4		

A

A*

b Plot the graph of $y = 2^x$ for $-3 \leqslant x \leqslant 4$ (Take y-axis from 0 to 20)

c Use your graph to estimate the value of y when $x = 2.5$

d Use your graph to estimate the value of x when $y = 0.75$

PS 6 Granny has two nephews, Alf and Bert. She writes a will leaving Alf £1000 in the first year after her death, £2000 in the second year after her death, £3000 the next year and so on for 20 years. She leaves Bert £1 the first year, £2 the second year, £4 the next year and so on.

a Show that the formula $500n(n + 1)$ gives the total amount that Alf gets after n years.

b Show that the formula $2^n - 1$ gives the total amount that Bert gets after n years.

c Complete the table for the total amount of money that Bert gets.

Year	2	4	6	8	10	12	14	16	18	20
Total	3	15	63							

d Draw a graph of both nephews' total over 20 years. Take the x-axis from 0 to 20 years and the y-axis from £0 to £110 000.

e Which nephew gets the better deal?

AU 7 A curve of the form $y = ab^x$ passes through the points (0, 3) and (2, 48). Work out the values of a and b.

Problem-solving Activity

Drawing linear graphs

a Draw the line $y = 2x + 3$ on a grid like the one shown here.

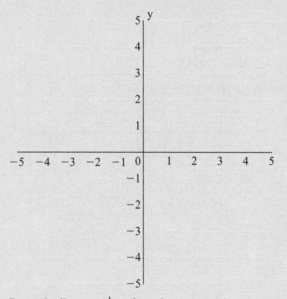

Draw the line $y = -\frac{1}{2}x + 3$ on the same grid.
What do you notice about the angle between the lines?

b Draw the lines $y = -4x + 1$ and $y = \frac{1}{4}x + 1$ on the grid.
What do you notice about the angle between the lines?

c Draw the lines $y = \frac{2}{3}x - 1$ and $y = -\frac{3}{2}x - 1$ on the grid.
What do you notice about the angle between the lines?

d Look at the equations of the lines in parts **a**, **b** and **c**.
What is the connection between the gradients in each part?

e Write down a rule for the gradients of perpendicular lines.

13.1 Circle theorems

1 Find the value of *x* in each of these circles with centre O.

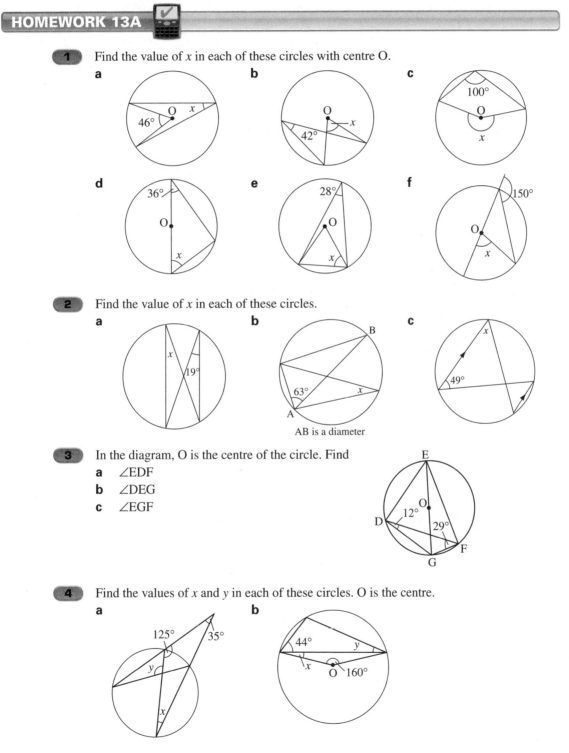

a

b

c

d

e

f

2 Find the value of *x* in each of these circles.

a

b

AB is a diameter

c

3 In the diagram, O is the centre of the circle. Find

a ∠EDF

b ∠DEG

c ∠EGF

4 Find the values of *x* and *y* in each of these circles. O is the centre.

a

b

A

AU 5 In each diagram, O is the centre of a circle.

    a Calculate the value of angle a.   **b** Calculate the value of angle b.

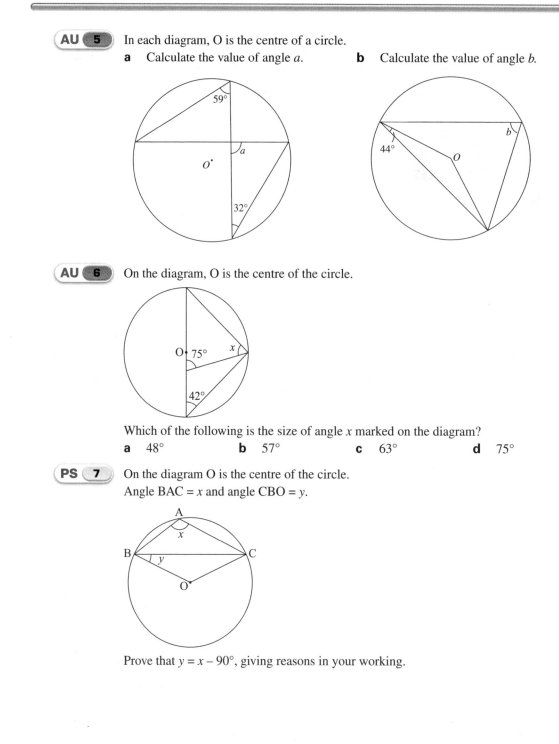

AU 6 On the diagram, O is the centre of the circle.

Which of the following is the size of angle x marked on the diagram?

  a 48°   **b** 57°   **c** 63°   **d** 75°

A*

PS 7 On the diagram O is the centre of the circle.

Angle BAC = x and angle CBO = y.

Prove that $y = x - 90°$, giving reasons in your working.

13.2 Cyclic quadrilaterals

HOMEWORK 13B

1 Find the size of the lettered angles in each of these circles.

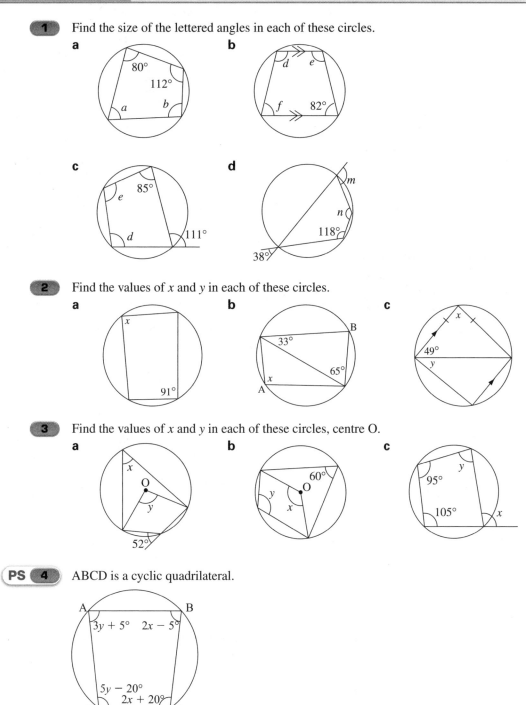

a

80°
112°
a   *b*

b

d   *e*
f   82°

c

e   85°
d   111°

d

m
n
118°
38°

2 Find the values of *x* and *y* in each of these circles.

a

x
91°

b

B
33°
65°
x
A

c

x
49°
y

3 Find the values of *x* and *y* in each of these circles, centre O.

a

x
O
y
52°

b

60°
O
y
x

c

y
95°
105°
x

PS 4 ABCD is a cyclic quadrilateral.

A   B
3*y* + 5°   2*x* − 5°
5*y* − 20°
2*x* + 20°
D   C

Work out the values of *x* and *y*.

A

AU **5** On the diagram, O is the centre of the circle.
Explain why the angle BOD is 128°.
Give reasons for your answer.

A*

6 ABCD are points on a circle. AB is parallel to CD.
Prove that $x = y$

13.3 Tangents and chords

HOMEWORK 13C

B

1 In each diagram, TP and TQ are tangents to a circle, centre O. Find values for r and x.

a

b

2 Each diagram shows a tangent to a circle, centre O. Find each value of y.

a

b

3 Each diagram shows a tangent to a circle, centre O. Find x and y in each case.

a

b

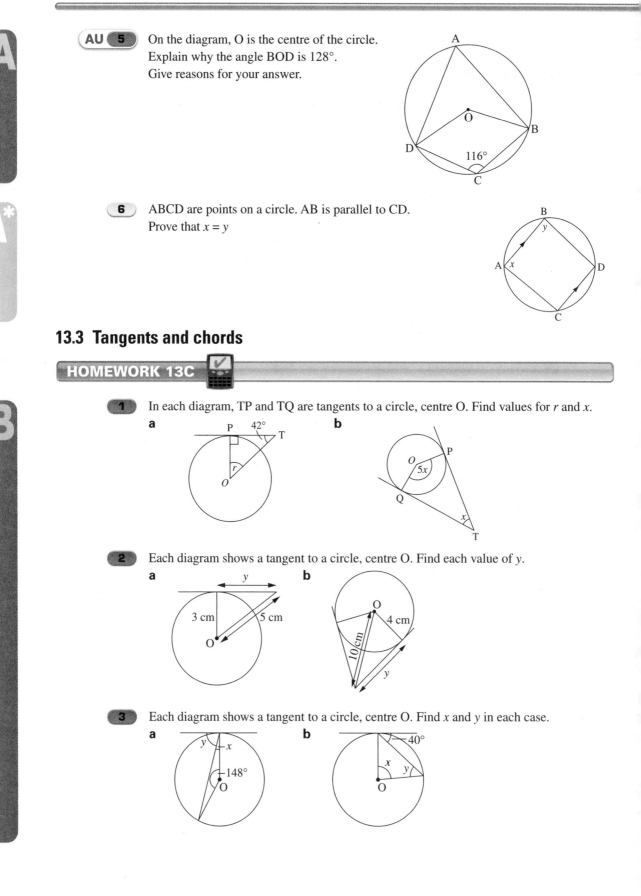

4 In each of the diagrams, TP and TQ are tangents to the circle, centre O. Find each value of *x*.

a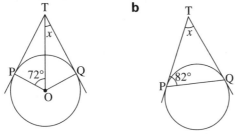

b

PS 5 The diagram shows two circles touching at X.

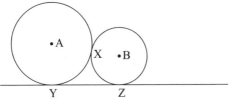

The circles have a common tangent at Y and Z.
The circle with centre A has a radius of 6 cm.
The circle with centre B has a radius of 3 cm.
Calculate the length YZ.

AU 6 Two circles intersect at X and Y.

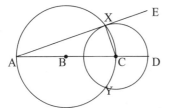

B and C are the centres of the circles and ABCD is a straight line.
Prove that the line AE is a tangent to the small circle.

13.4 Alternate segment theorem

HOMEWORK 13D

1 Find the size of each lettered angle.

a

b

2 In each diagram, find the value of x.

a

b

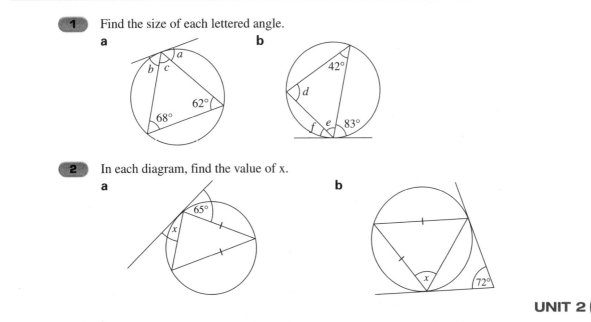

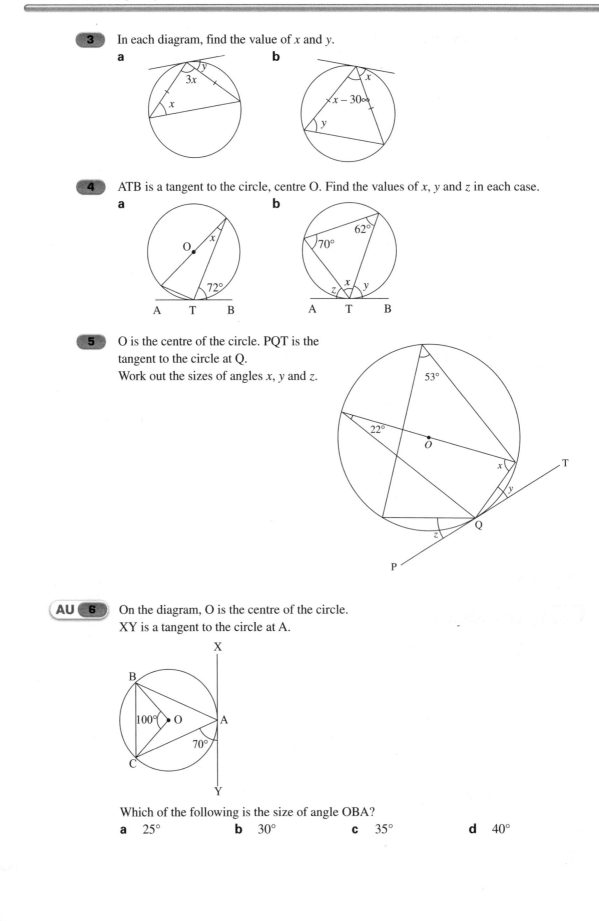

3 In each diagram, find the value of x and y.

a

y
$3x$
x

b

x
$x - 30°$
y

4 ATB is a tangent to the circle, centre O. Find the values of x, y and z in each case.

a

O
x
72°
A   T   B

b

62°
70°
x
z y
A   T   B

5 O is the centre of the circle. PQT is the tangent to the circle at Q.

Work out the sizes of angles x, y and z.

53°
22°
O
x
T
y
Q
z
P

AU 6 On the diagram, O is the centre of the circle.

XY is a tangent to the circle at A.

X
B
100°   • O   A
70°
C
Y

Which of the following is the size of angle OBA?

a   25°   **b**   30°   **c**   35°   **d**   40°

PS 7 AB is a tangent to the circle at X.
YZC is a straight line.

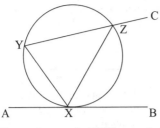

Prove that angle BXY = angle XZC.

Functional Maths Activity

Security cameras

A museum has a circular room which is used to show some of its valuable exhibits.
The room needs to be completely covered by security cameras placed around the wall of the room.
The curator of the museum wants to use as few cameras as possible to keep the cost low and to make the supervision easier.
The curator is thinking of using security cameras with an angle of view of 60°.

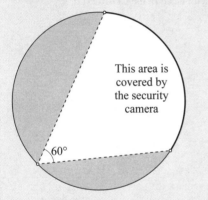

1   How many cameras will the curator need for the room?
2   How should the cameras be arranged?

William Collins' dream of knowledge for all began with the publication of his first book in 1819. A self-educated mill worker, he not only enriched millions of lives, but also founded a flourishing publishing house. Today, staying true to this spirit, Collins books are packed with inspiration, innovation and practical expertise. They place you at the centre of a world of possibility and give you exactly what you need to explore it.

Collins. Freedom to teach.

Published by Collins
An imprint of HarperCollins*Publishers*
77–85 Fulham Palace Road
Hammersmith
London
W6 8JB

Browse the complete Collins catalogue at
www.collinseducation.com

10 9 8 7 6 5 4 3 2 1

ISBN-13 978-0-00-733995-2

Brian Speed, Keith Gordon, Keith Evans, Trevor Senior and Chris Pearce assert their moral rights to be identified as the authors of this work

British Library Cataloguing in Publication Data
A Catalogue record for this publication is available from the British Library

Commissioned by Katie Sergeant
Project managed by Patricia Briggs
Edited by Brian Asbury
Answers checked by Steven Matchett and Joan Miller
Cover design by Angela English
Concept design by Nigel Jordan
Illustrations by Wearset Publishing Services
Typesetting by Wearset Publishing Services
Production by Leonie Kellman
Printed and bound by L.E.G.O. S.p.A. Italy

Important information about the Student Book CD-ROM
The accompanying CD-ROM is for home use only. You cannot copy or save the files to your hard drive and it will work only when placed in the CD-ROM drive.

NOTES

NOTES

NOTES

NOTES

NOTES

NOTES

NOTES